6th NanoInnovation Conference and Exhibition (NanoInnovation 2021)

IOP Conference Series: Materials Science and Engineering
Volume 1265

Rome, Italy
21 – 24 September 2021

ISBN: 978-1-7138-6425-7
ISSN: 1757-8981

Printed from e-media with permission by:

Curran Associates, Inc.
57 Morehouse Lane
Red Hook, NY 12571

Some format issues inherent in the e-media version may also appear in this print version.

Institute of Physics
Dirac House, Temple Back
Bristol BS1 6BE UK

Phone: 44 1 17 929 7481
Fax: 44 1 17 920 0979

techtracking@iop.org

Additional copies of this publication are available from:

Curran Associates, Inc.
57 Morehouse Lane
Red Hook, NY 12571 USA
Phone: 845-758-0400
Fax: 845-758-2633
Email: curran@proceedings.com
Web: www.proceedings.com

TABLE OF CONTENTS

IOP Conf. Series: Materials Science and Engineering 1265 (2022) 011001 doi:10.1088/1757-899X/1265/1/011001

Preface - NanoInnovation 2021

The 6th edition of the NanoInnovation, Conference and Exhibition, has been held September 21-24, 2021, hosted as usual in the Renaissance Cloister by Giuliano da Sangallo of the Faculty of Civil and Industrial Engineering of SAPIENZA University of Rome (Italy). The peculiar situation of the COVID-19 pandemic, showing a momentaneous break in the epidemy, allowed to conduct the conference in presence even if, to favor a larger audience and to support participants with difficulty in travelling, all the events were available also by remote connection.

NanoInnovation is the main Italian meeting place for the wide and multidisciplinary community involved in the development of nanotechnologies and their integration with other key enabling technologies (KETs), in all application fields. The event aims at involving the scientific, industrial, and business communities in an open discussion environment, where researchers, technologists, managers, entrepreneurs and investors can share experiences, opinions and expectations about the role of nanotechnologies for future innovations.

As in the already established protocol the Conference, with a strong bottom-up character, was co-organized with the support of the major Italian players in nanotechnology coming both from the public and academic research and from the industrial sector.

In this edition, 24 co-organizers contributed to the success of the initiative by supporting the organization and by proposing interesting and timely themes where to concentrate the presentations and the debate. The conference, which was held for the first time in a hybrid with speakers and attendant both in presence and remotely connected, has attracted fellow researchers, academicians, and industry professionals recording more than 1500 registered participant with 600 attendants in presence. The dense programme of NanoInnovation 2021 was managed by 80 chairs and included the talks by more than 500 speakers, 1 opening session, 1 round table, 2 scientific plenary sessions, 4 parallel scientific lectures, 1 technical lecture, 1 *lectio magistralis* held by Dr. Venkatraman Ramakrishnan, 2009 Nobel Prize in Chemistry, on "My adventures in the ribosome: Nature's amazing nanomachine", 79 technical thematic symposia, 6 scientific multi-sessions workshops, 4 special events, 5 joint events, 2 schools, 1 update training organized in collaboration with Ordine degli Ingegneri della Provincia di Roma, 8 satellite events, 84 posters in a dedicated session and 1 NEST prize for young researchers. NanoInnovation 2021 could rely on 20 supporters and 12 institutional and scientific patronages.

A technical commercial exhibition hosting 21 companies presenting their products and technical solutions has been also organized.

The present special issue is a collection of selected contributions presented during NanoInnovation 2021 - Conference and Exhibition. The papers concern different topics discussed during the conference, representing the state of the art and the outlook of nanoscience and related technology in the fields of Agri-Food, Health & Nanomedicine, Energy & Environment, Microscopy, Characterization & Imaging.

Conference Chair:

- Marco ROSSI - Sapienza University of Rome

Organizing Committee:

- Marco ROSSI - Sapienza University of Rome
- Massimo BERSANI – FBK
- Marco CRESCENZI – ISS
- Vittorio MORANDI - CNR IMM
- Fabrizio PIRRI - Polytechnic University of Turin
- Andrea PORCARI - AIRI

Steering Committee:

- Sesto VITICOLI – AIRI
- Piero BAGLIONI - University of Florence & CSGI
- Giorgio Mariano BALESTRA - University of Tuscia

- Flavia BARONE – ISS
- Marzia BEDONI – Don Gnocchi Foundation
- Claudia L. BIANCHI – University of Milan
- Luca BOARINO – INRIM
- Fabio BOCCUNI – INAIL
- Chiara CAPPELLI – Scuola Normale Superiore
- Ennio CAPRIA – European Synchrotron (ESRF)
- Giuliano CASATI – Assing
- Gianluigi CASSE – FBK
- Sabrina CONOCI – University of Messina / CNR-IMM / distretto tecnologico Sicilia Micro e Nano Sistemi
- Filippo D'ARPA – Distretto tecnologico Sicilia Micro e Nano Sistemi
- Marco DI CLEMENTE – ASI
- Giuliana FAGGIO - The "Mediterranean" University of Reggio Calabria
- Lorenza FERRARIO – FBK
- Guglielmo FORTUNATO – CNR - IMM
- Narciso GAMBACORTI – CEA - Leti
- Roberto GIANNANTONIO – CNR - IRSA
- Paolo GRIANTI – JEOL ITALIA
- Andrea LAMBERTI – Polytechnic University of Turin
- Giulio LAMEDICA – ZEISS
- Claudio LANZIERI – Leonardo
- Nicola LISI – ENEA
- Luca MARCHIOL – University of Udine
- Francesco MATTEUCCI – European Innovation Council (EIC)
- Donatella PAOLINO - University "Magna Graecia" of Catanzaro
- Giuseppe ROSACE – University of Bergamo
- Maria Sabrina SARTO – Sapienza University of Rome
- Giovanni TOSI - University of Modena and Reggio Emilia

Programme Committee:

- Marco VITTORI ANTISARI - NanoItaly Association
- Fabrizio BARBERIS - University of Genoa
- Francesco BIANCARDI – ZEISS
- Stefano BIANCO - Polytechnic University of Turin
- Gabriele BULLA - JEOL ITALIA
- Annarica CALCABRINI – ISS
- Sebastiano CAMPAGNA - University of Messina
- Andrea CANESCHI - INSTM
- Ennio CAPRIA - ESRF
- Maria CARAFA – Sapienza University of Rome
- Filippo CASCONE – Register of Engineers / Foundation Register of Engineers
- Gabriele CENTI – University of Messina
- Massimo CERRI – Register of Engineers
- Annalisa CHIAPPONE – Polytechnic University of Turin
- Luigi Carlo CHIARENZA - Foundation Register of Engineers
- Giancarlo CICERO - Polytechnic University of Turin
- Isabella DE ANGELIS -ISS
- Luciana DINI – Sapienza University of Rome
- Marta FEROCI - Sapienza University of Rome
- Andrea GAIARDO – FBK
- Ettore GAROFALO – Assing
- Rodorico GIORGI – University of Florence / CSGI
- Sergio IAVICOLI – INAIL
- Leonardo MATTIELLO – Sapienza University of Rome
- Stefania MESCHINI – ISS
- Giacomo MESSINA - The "Mediterranean" University of Reggio Calabria

IOP Conf. Series: Materials Science and Engineering 1265 (2022) 011001 doi:10.1088/1757-899X/1265/1/011001

- Agnese MOLINARI – ISS
- Giovanni NICOLAI – Register of Engineers
- Siglinda PERATHONER – University of Messina
- Donatella PIETRAFORTE – ISS
- Pasqualantonio PINGUE – Scuola Normale Superiore
- Maria Elena PISANU – ISS
- Georg PUCKER – FBK
- Giuliano RANUZZI – ASI
- Ignazio ROPPOLO- Polytechnic University of Turin
- Stefania SABELLA – IIT
- Daniele SCHIAVI – University of Tuscia
- Annarita STRINGARO – ISS

Honorary Committee:

- Luigi AMBROSIO - Scuola Normale Superiore
- Don Vincenzo BARBANTE - Don Gnocchi Foundation
- Franco BETTONI – INAIL
- Silvio BRUSAFERRO – ISS
- Maria Chiara CARROZZA – CNR
- Salvatore CUZZOCREA - University of Messina
- Giovambattista DE SARRO - University Magna Graecia of Catanzaro
- Gabriele GALATERI DI GENOLA – IIT
- Luigi NICOLAIS - Ministero dell'Università e della Ricerca
- Francesco PRIOLO - University of Catania
- Francesco PROFUMO - Bruno Kessler Foundation
- Francesco SETTE – ESRF
- Santo Marcello ZIMBONE - The "Mediterranean" University of Reggio Calabria
- Diederik Sybolt WIERSMA – INRIM

The Editors of the Proceedings of NanoInnovation 2021 are deeply grateful to all the Authors who contributed to the success of both the conference and the present issue, and to the Reviewers for their professional support, competence, carefulness, and patience.

The Editors

Marco ROSSI – Sapienza University of Rome
Marco VITTORI ANTISARI – NanoItaly Association
Daniele PASSERI – Sapienza University of Rome

IOP Conf. Series: Materials Science and Engineering 1265 (2022) 011002 doi:10.1088/1757-899X/1265/1/011002

Peer Review Statement

All papers published in this volume have been reviewed through processes administered by the Editors. Reviews were conducted by expert referees to the professional and scientific standards expected of a proceedings journal published by IOP Publishing Publishing.

- **Type of peer review:** Single Anonymous
- **Conference submission management system:** Morressier
- **Number of submissions received:** 15
- **Number of submissions sent for review:** 16
- **Number of submissions accepted:** 14
- **Acceptance Rate (Submissions Accepted / Submissions Received × 100):** 93.3
- **Average number of reviews per paper:** 1.1
- **Total number of reviewers involved:** 15
- **Contact person for queries:**
  Name: Daniele Passeri
  Email: daniele.passeri@uniroma1.it
  Affiliation: University of Rome Sapienza, Department of Basic and Applied Sciences for Engineering

The role of hydrogen from electrolysis in the overproduction of energy fromrenewable sources

A Pozio and S Galli

ENEA CR Casaccia, TERIN-PSU-ABI, Via Anguillarese 301, 00123, Rome, Italy

Abstract. The annual production from renewable electricity sources in Italy has been considerably growing and in some areas, in particular Southern Italy, there is constantly an overproduction of electricity during the peak hours of insulation due to the photovoltaic systems. Such excess electric energy could be used to produce large amount of hydrogen by electrolysis, thus covering approximately 3.3% of the yearly national industrial hydrogen demand and avoiding emissions of approximately 85,000 tons of carbon dioxide.This work shows that few large alkaline electrolysis plants, located in strategic sites and supplied with this excess energy, can just now produce hydrogen for industrial use at competitive costs. Furthermore, the presence of electrolysis plants connected to the main electricity grid would represent a controllable element useful for its regulation and stability.

1. The renewable energy production in Italy

For over a decade, Italy has been witnessing a continuous growth in the installation of energy production plants from renewable sources. In figure 1, which shows the variation in gross installed power over the last eighty years by type of plant, we noticed (a) a general increase in the overall value, due to a progressive increase in consumption and the industrialization of the country and (b) the substantial increase, in the last twenty years, in the share of renewable energy plants, in particular wind and photovoltaic, to the detriment of energy produced from fossil sources.

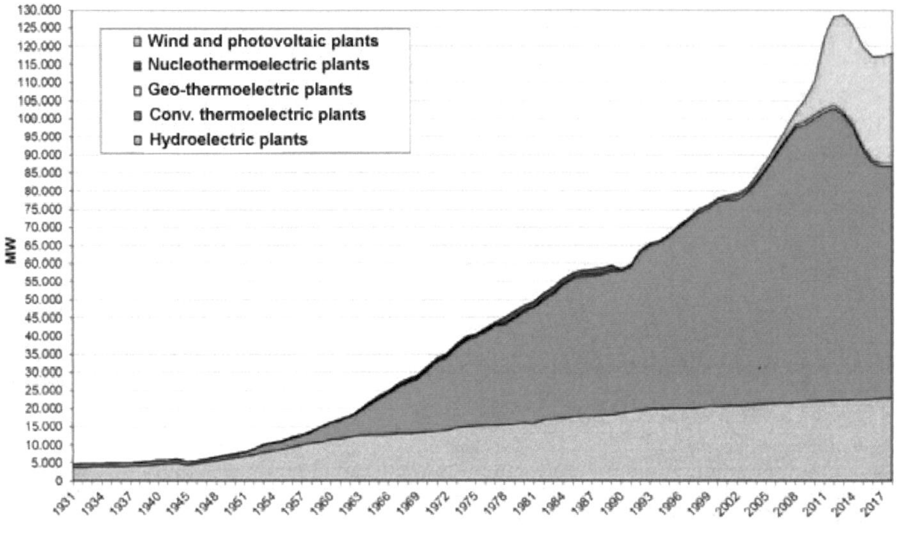

Figure 1. Gross efficient power installed in Italy from 1931 to today [1].

Figure 2 details how the overall national energy demand is distributed among the various type of primary resource. Energy consumption in Italy has been diversified considerably in the last decade, depending less on oil againsta considerable growth of the energy production from renewable energy sectors:in 2019 against an installed power of 119.3 GW and an overall energy consumption of 294 TWh, the contribution from renewable energy was approximately 114.4 TWh, equal to 38.9%. A substantial part of thisenergy is related to the so-called renewable electricity (wind and photovoltaic, ER_{el}) which, with their 42 TWh of consumption, corresponds to about 14.2% of the total energy national production.

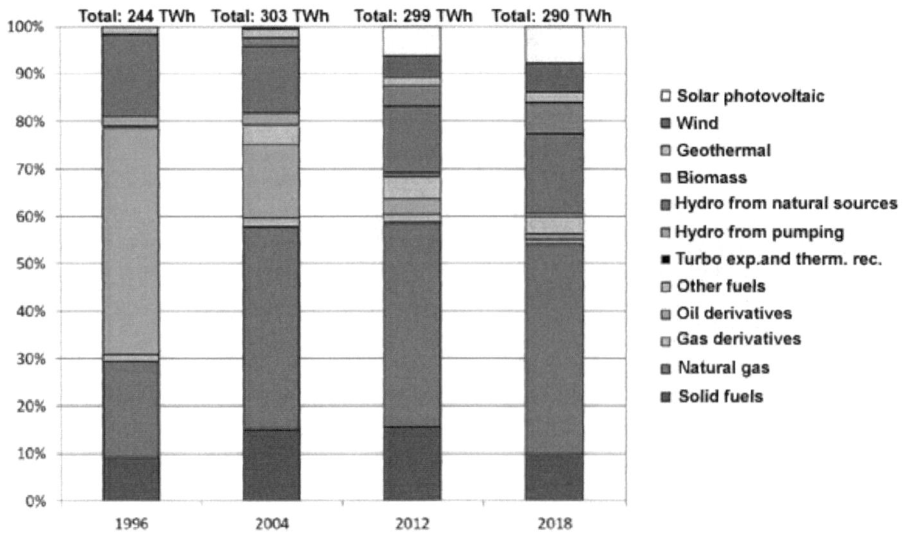

Figure 2. Variation of the energy production mix [1].

In order to achieve the 2030 objectives indicated by the Integrated National Energy and Climate Plan (PNIEC) [2], an increase of about 39 GW in ER_{el} sources will be added at the current installed one (31.6 GW - 2019), so reaching a total of 70.5 GW (2030), a value of ER_{el} power more than double of the present one.

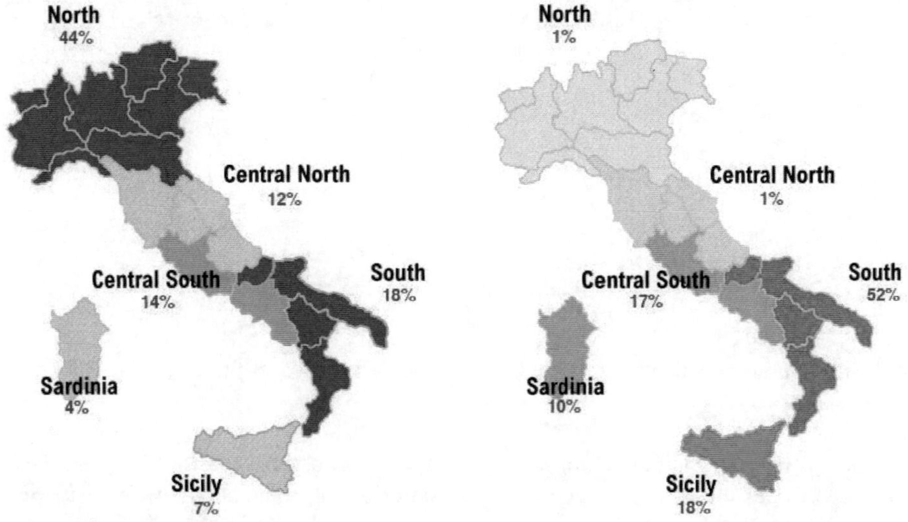

Figure 3. Photovoltaic (blue) and wind (green) power installed in Italy by market area in 2018 [3].

In Italy energy production from wind and photovoltaic power plants is not always homogeneously distributed (figure 3): if the energy produced by photovoltaics in the North (10.2 TWh) is quite similar to that produced in the South (10.1 TWh) (but with different installed power), the share of wind energy produced in the North is instead significantly lower (0.26 GWh vs 18.1 TWh) [3].

This imbalance in the production of energy from renewable sources, typical of other European countries, would not have great consequences if this energy were consumed and managed in the respective production areas. In Italy, even if globally the production of energy from renewable electricity is generally always lower than the daily demand for electricity (figure 4), in specific areas this can't be true. So while in the North the daily demand for electricity can absorbregularlyall production from renewable, in the South, in the same average day and in the few central daytime hours, this production of energy can exceeds the electricity load demand (figure 4).

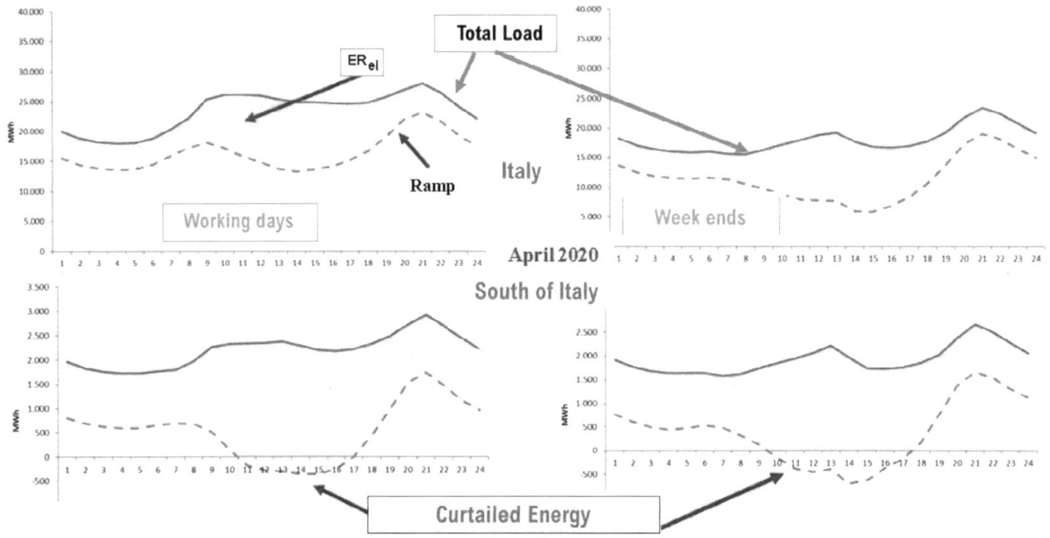

Figure 4. Daily average profiles of energy consumption in Italy and the South in working days and weekends [1].

The local energy overproduction from renewable energy (due mainly to PV plants in the peak insulation hours) forces the national electric grid operator to act, transferring as much as possible this excess energy on the distribution and transmission network and, eventually, by disconnecting part of the renewable generation plants (generally Wind energy). This amount of electricity not absorbed annually by the grid (MPE: Mancata Produzione Eolica – *in italian* - Wind Production Lack) exhibits a growing trend (fig. 5), and looks to be greater with the expected increase in share of renewable electricity. Furthermore the electricity grid operator generally has to pay the producer a fee, currently very low (0.04 Euro/MWh in 2018) [4], to compensate for the lack of production and consequent sale of wind energy.

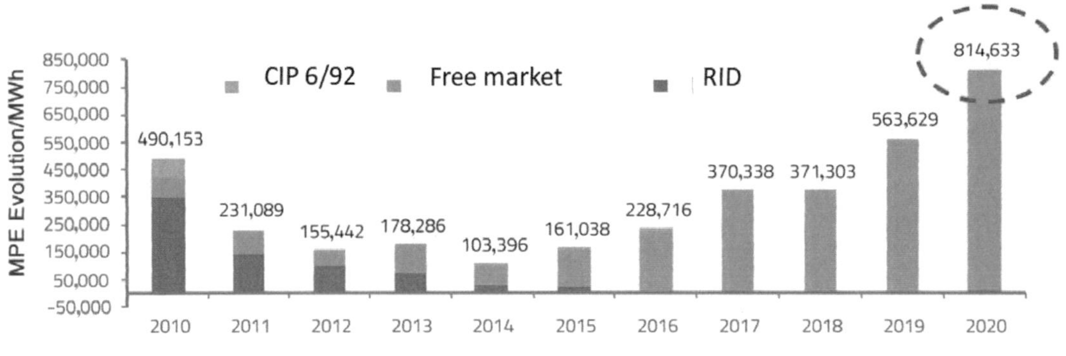

Figure 5. Evolution of relative energy to the MPE divided by commercial regime [5].

2. Overproduction of energy and demand for hydrogen

Considering this regional energy overproduction and with the aim of evaluating a possible solution for its profitable use, we hypothesize to produce hydrogen from electrolysis using this excess energy (figure 6).

The concept, certainly not a novelty in the integration studies of small or medium electrolysis plants in combination with renewable energy, consists in diverting this excess energy present in the electricity grid to electrolysis plants for the production of hydrogen [6].
Over the past decades a considerable number of this type of plant has been studied and demonstrate, in different sizes and with various technologies and control logics, then demonstrating their technical feasibility but also the strong dependence of the economy of these plants on the availability of low-cost electricity [7]. One of the driving reasons behind the hypothesis presented here is just the availability of a surplus energy and the possibility of conceiving the electrolysis system itself as a controllable element of the national electrical network, capable to absorb it and at the same time useful for by battery systems [8] and abroad on electric grid for frequency control [9], support this hypothesis. its regulation and stability. Similar exploratory experiences, conducted in Italy on electric grid.

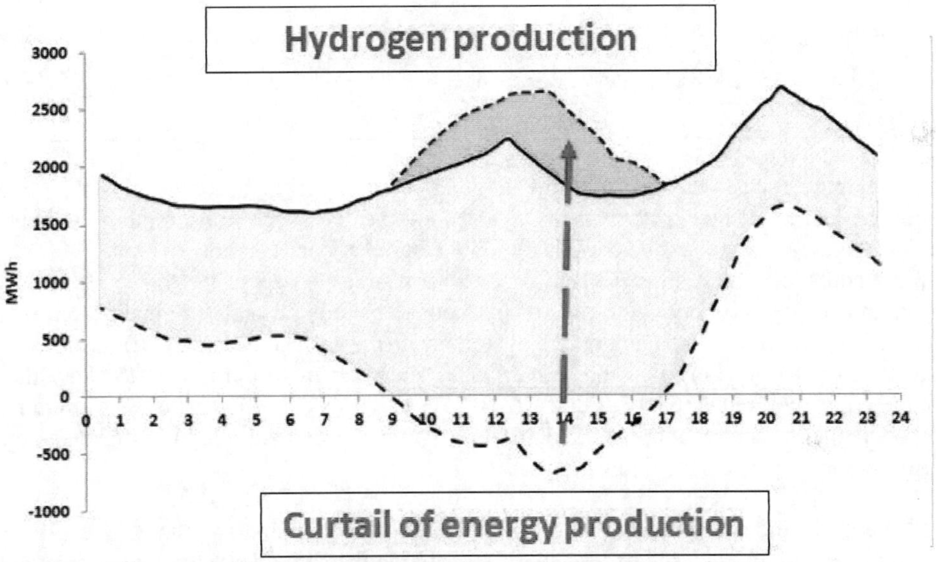

Figure 6. Excess daily renewable energy profile can be used to produce hydrogen.

To have a rough estimate of hydrogen production with the current excess energy of the network, and its impact on industrial consumption in Italy, an electrolysis efficiency of 65% was considered in line with the current values indicated by this type of plant.
According to official electric grid data [5], the lack of energy injected by renewable electricity plants (mostly Wind energy) amounts to approximately 815 GWh in 2020, whose distribution in the southern regions and islands is shown in table 1, estimating it divided according to the share of renewable electricity supplied by each.

Table 1. Distribution of energy consumption in Southern Italy from renewable electricity sources [10].

	Wind/GWh	PV/GWh	ER_{el}/GWh	Regional prod/%	Total GWh/year
Molise	662	231.2	893.2	3.2	25.7
Abruzzi	410,2	945.5	1355.7	4.8	39.0
Calabria	2132.4	681.3	2813.7	9.9	81.0
Sardinia	1677.1	1154.7	2831.8	10.0	81.5
Basilicata	2423	491.3	2914.3	10.3	83.9
Campania	3209.2	981.5	4190.7	14.8	120.6
Sicily	2765.4	1911.3	4676.7	16.5	134.6
Apulia	4801.9	3839.2	8641.1	30.5	248.7
TOTAL	18081.2	10236.0	28317.2	100.0	815.0

With the current excess energy from the national electric grid approximately 16,000 tons/y of hydrogen could be produced, equivalent to 3.3% of the hydrogen consumption in industry, that amounts to about 480,000 tons/y (16 TWh/y - LHV 120 MJ/Kg) [11].
With the amount of excess energy from renewable sources produced in the various regions, and assuming an average daily operating time of 6.5 hours/day, it is possible to obtain the total power of electrolysers required for each region to absorb such a surplus of energy (figure 7 left). These hypothesized values have only an indicative value, and their main purpose is to show the order of magnitude of the required electrolysis units. A 10 MW electrolysis plant was used as a standard unit as already operating in other countries [12,13,14], and as it can provide a roughly estimate of the number of electrolysis units per region (figure 7 right).

On the basis of the increase in power of the ER_{el} plants foreseen in 2030 by the PNIEC (70.5 GW) and assuming conservatively that the excess energy maintains the same percentage of current proportions with respect to consumption (1.95%), it is possible to calculate the future yearly excess (1,817 GWh) and therefore, estimate the production of hydrogen at that date which would be around 35,000 tons/y, equivalent to 7.7% of the current consumption of hydrogen, still a limited fraction of the total consumption. This forecast estimate assumes that electricity consumption remains unchanged.

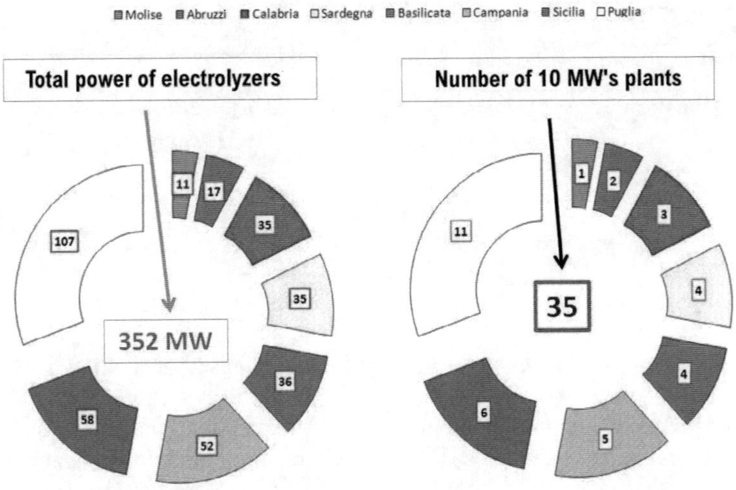

Figure 7. Estimation of the regional amount and distribution of standard 10 MW electrolysis units to be installed to absorb present local excess ER_{el}.

The current hydrogen market is aimed at four main sectors: chemistry, refining, metal working and diversified industrial applications [15]. The largest quantity is consumed for over 90% in the first two sectors, while the remainder (about 48,000 tons/y) is produced in specific plants and supplied in cylinders, tankers, as gas and liquid, to various industrial sectors (e.g. steel mills, food industry, semiconductors, glass, etc.). These latter sectors could be the first recipients of electrolytically produced hydrogen in terms of quality, quantity and distribution methods.

Presently, the hydrogen market is diversified, and hydrogen producers are often self-consumers, as they produce large quantities of hydrogen exclusively for their own internal industrial consumption (ammonia, methanol, resins). There are also manufacturers on behalf of third parties, able to resell variable quantities of hydrogen produced in dedicated plants, and sometimes these plants are part of the same industrial areas or are connected to specific industries. Finally, there are also industrial sectors that produce hydrogen as a by-product of other processes (e.g. chlorine production plants). The boundary between these subjects is not always defined (figure 8): self-producers (1) can become suppliers of the retailers themselves in case of overproduction (2) or consumers in case of need (3) as well as producers from other processes they can become suppliers of the former (5-7) or of the latter (5-6).

The location of the plants in Italy (figure 9) shows a distribution over the entire national territory with a certain concentration in the industrial areas of the North-West of the country but also in the South, in particular Puglia and Sicily.

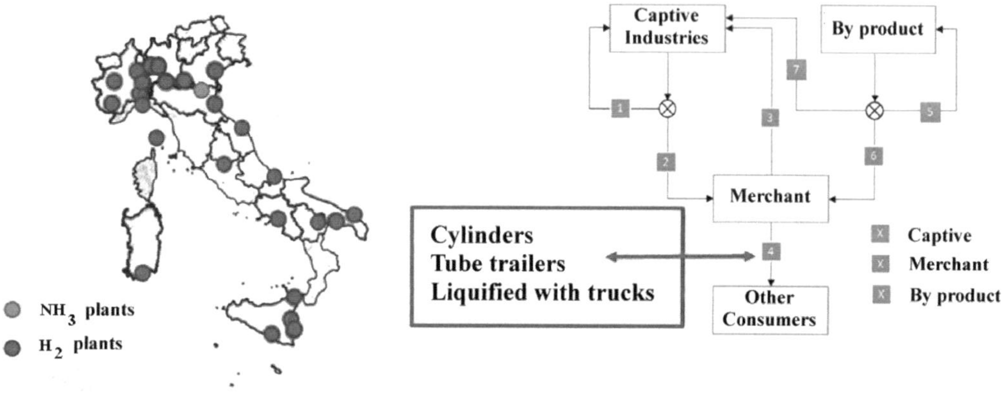

Figure 8. Definition of types of hydrogen production based on availability [15].

Once the order of magnitude of the various regional plants necessary to absorb the surplus energy has been established, the next step could consist in identifying their location. Here it is necessary to introduce considerations that take into account various factors that are with priority in charge of the national electric and gas grid operators (figure 9), whose main objectives are to ensure the stability and the quality of the respective service.

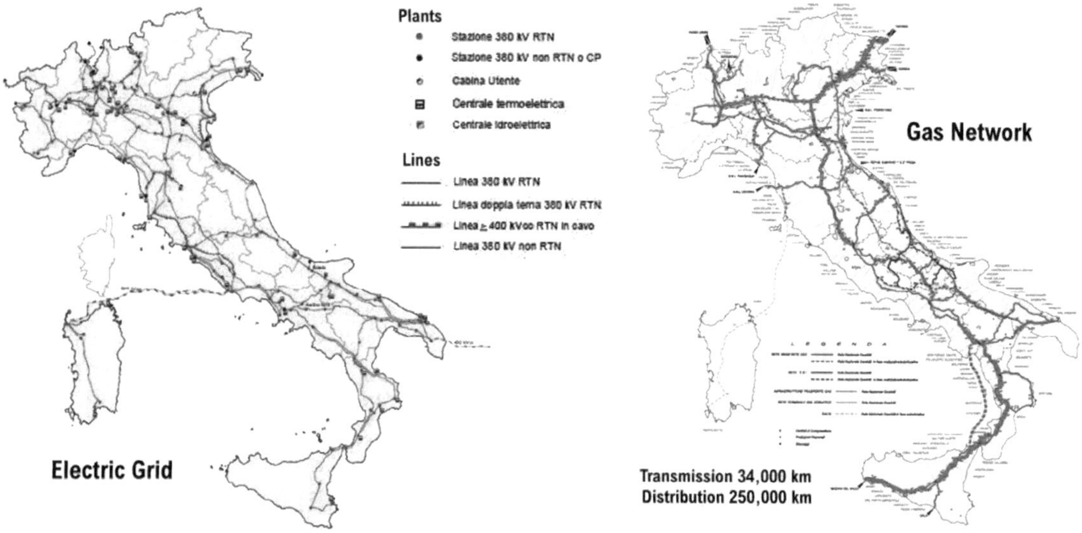

Figure 9. Configuration of the electricity network (left) and gas line network (right) in Italy.

A series of criteria can be indicated to identify potential sites:

- congestion points of the Medium Voltage electricity network
- presence of industrial activities for a direct use of pure hydrogen
- proximity to the gas network for possible injection (P2G - Power to Gas)

The large, fast and almost continuous adjustable load of an electrolysis system could represented an effective and rapid element to be placed in critical sites along the electrical network for the electric energy balance.

It is clear that a large number of sites, while allowing better distributed network control, would have higher investment and management costs. A concentration of several electrolysis modules in few sites would benefit of economy of scale, due to less critical auxiliary systems (inverters, rectifiers, water and gas purification systems, compressors, control systems, battery, etc.) and for a greater flexibility in optimal management of the system and maintenance of the individual units.
Industrial end users in the vicinity of the plant itself would be favorable, so as to minimize storage, compression and transport systems; the direct and immediate use of the high purity hydrogen (oxygen as well) produced would be ideal.
The presence of the gas network for a possible injection into the network, even if it does not appear to be the optimal solution from a chemical and thermodynamic point of view, could be still an enough simple and economical solution to implement in the future, once the concentration constraints and piping materials issues have been overcome.
These large electrolytic hydrogen production plants directly connected to the electric grid could be substantially standardized and modular, easily expandable according to the requests for increased power and needs of use. Compared to the various demonstration electrolysis plants connected with wind or photovoltaic systems and built up to now, the specifications of such an industrial electrolysis systems look more standardized and economically competitive, having specific and more constant electrical characteristics as input energy data and consequently optimized and standardized electrical, power electronic, electrochemical and auxiliaries components (stack, deoxo, reverse osmosis, compression, AC/DC and DC / DC converter, sensors, etc.).

3. Comparing technologies: alkaline and polymer electrolysers

The low-temperature electrolysis technologies currently mature to meet the needs highlighted above in relation to the quantities of hydrogen to be produced and the power required are essentially two: alkaline electrolysers and polymeric membrane electrolysers (PEM: Polymer Electrolyte Membrane). Alkaline, commercially more widespread, are generally larger in size and have lower costs due to the absence of precious metals group electrocatalysts (PGM: Platinum Group Metal). Table 2 compares the technical characteristics of two electrolysis stacks of similar power (AC-485 alkaline and polymeric PEM-MC-400), both manufactured by NEL (Norway), an industry with almost a century of experience in electrolysis technologies.
The choice of similar units in production capacity (413 for alkaline and 485 Nm^3/h for polymeric one) makes the technical characteristics and performances comparable. By coupling multiple stacks, larger plants can be created, e.g. the alkaline A3380 (2.1 MW) consisting of 8 AC-485 modules and the polymeric M4000 (1.8 MW) composed of 10 MC-400 modules.

Table 2. Comparison table of alkaline and polymer electrolyser [16].

Parameters	NEL AC485 (262 kW)	NEL PEM-MC400 (180 kW)
Net production rate	485 Nm^3/h	413 Nm^3/h
Prod Capacity dynamic range	15%...100% of flow range	10%...100% of flow range
Delivery pressure	1-30 bar (g) 200 bar with additional compression	1-30 bar (g) 200 bar with additional compression
Spec. power consumption	4.40 kWh/Nm^3	4.53 kWh/Nm^3
System electrical efficiency	67.7 %	65.7 %
Hydrogen purity	99.6 % (99.99 with purifier)	99.9998 %
Electrolyte	25% KOH solution	PEM
Feed water consumption	0.9 l/Nm^3	0.9 l/Nm^3 (H_2O a 0.1 $\mu S\ cm^{-1}$)
Footprint	225 m^2	160 m^2
Stack life time	> 10 years	Not verified
Catalyst – Material	No PMG	PMG (Ir,Pt) – Ti, Nafion©

The modularity of the electrolysis systems for both technologies with a stack size limit simplifies a series of problems such as pressure seal, fluid dynamics (two-phase electrolyte gas mixture), maintenance and the electrical supply to the entire plant.

The ability to operate at partial load, a condition that should be typical when the electrolysis plant is powered by renewable energy, is broad for both technologies, more limited for the alkaline electrolyser given the greater diffusion of gases through the membrane with the consequent risk of formation of explosive mixtures, especially in the stand-by phases. It should in any case be considered that, with the presence of multiple electrolysers modules and with careful management of the energy entering the production site, the limit operating conditions could be kept under control.

A similar argument can be applied to the production of hydrogen under pressure, where an initial level of pressurization could be implemented directly inside the electrolyser itself, thus reducing the subsequent compression stages. Although potentially more adapt for high pressure operation, the long-term performance of the polymer electrolyser still needs to be verified in the field in high power units, when periodic and repetitive stack pressurization could produce pernicious relaxations in the sealing parts. At the moment a compression system downstream of the electrolysers, with adjustable pressure levels, adapted to the requested service of hydrogen, looks to be the most practicable, albeit probably less economical solution.

Both technologies have no differences in the electrical power supply system as well as in the auxiliaries required for the treatment of the gas produced (elimination of impurities and drying), with less work required for the PEM technology given the better quality of the gas at the outlet. Not so for the process water supply, where the very low conductivity levels required by PEM technology (~ 0.1 $\mu S\ cm^{-1}$), to avoid membrane contamination, lead to the adoption of additional purification technologies (osmosis systems). The membrane contamination requires a continuous control of the conductivity of the circulating electrolyte and its constant purification, in order to eliminate the traces of ions released by the materials making up the circuit. Membrane contamination poses a great risk to the reliability and performance over time of polymer electrolysers. The footprint of the plant favors PEM technology, with more compact and smaller electrolysers, although, being industrial plants, this aspect appears to be of lesser importance.

Regarding the quality of the circuit materials and catalysts, if for the alkaline electrolysers the aggressiveness of the electrolyte (sodium and potassium hydroxide) leads to the use of adequate components in the circuitry, in the polymeric ones it is the stack section (electrodes, membrane and plates) that requires valuable and very high-quality and expensive materials (Pt, Ir, Titanium, Nafion®). In particular, the availability and restricted localization of PGMs will certainly limit the extensive use of this technology [17].

From a technical point of view, the alkaline technology appears more adequate and reliable than that of the PEM also by virtue of the fact that these plants have been in operation for many years; as an example, here it is recalled the case of the Norsk-Hydro (now NEL) electrolysis plant, capable of produce 70 ton/day (approx. 150 MW), and operated continuously from 1948 to 1990 (figure 10).

Figure 10. Norsk-Hydro 30,000 Nm^3/h hydrogen electrolysis plant.

4. Economic evaluation of hydrogen production from excess energy

There are many analysis on the cost of producing hydrogen by electrolysis and in general they all show that the cost of electricity as an incisive factor (approx. 60%) [18]. Starting from the specific consumption of the electrolyser (table 2) it is possible to easily construct the theoretical line of the cost of hydrogen as a function of the electricity price alone (0.08 €/kWh industrial price in Italy in 2021) [19], with the slope representing the energy consumption per unit of weight of hydrogen produced. Figure 11 shows the comparison between the lines related to the two models described above (AC-485 with 49.3 kWh/kg of H_2 and PEM M-400 50.7 kWh/kg of H_2) also highlighting the lower theoretical limit of the production cost, assuming an ideal electrolyser (39 kWh/kg @100% efficiency).

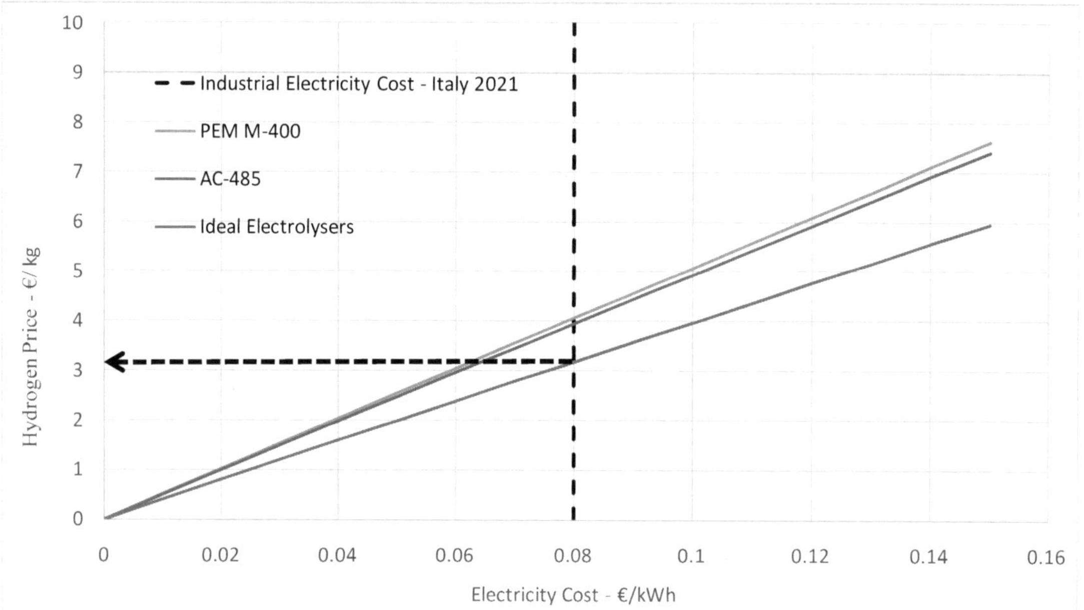

Figure 11. Cost of hydrogen produced by electrolysis (w/o CAPEX) for an ideal system and for the commercial electrolysers of the NEL PEM-M-400 and AC-485.

From figure 11, it can be observed that, with noplant costs (investment, maintenance, disposal, etc.), at the current price of electricity in Italy, the cost of hydrogen cannot fall below 3 €/kg, even using an ideal electrolyser. Obviously, commercial electrolysers have a higher hydrogen production cost (~ 4 €/kg).

A complete analysis, introducing the plant costs, compares alkaline and polymeric electrolysers with the option, for the alkaline one, of using products manufactured in Europe and in China. The additional costs were extracted from the Bloomberg report [20].

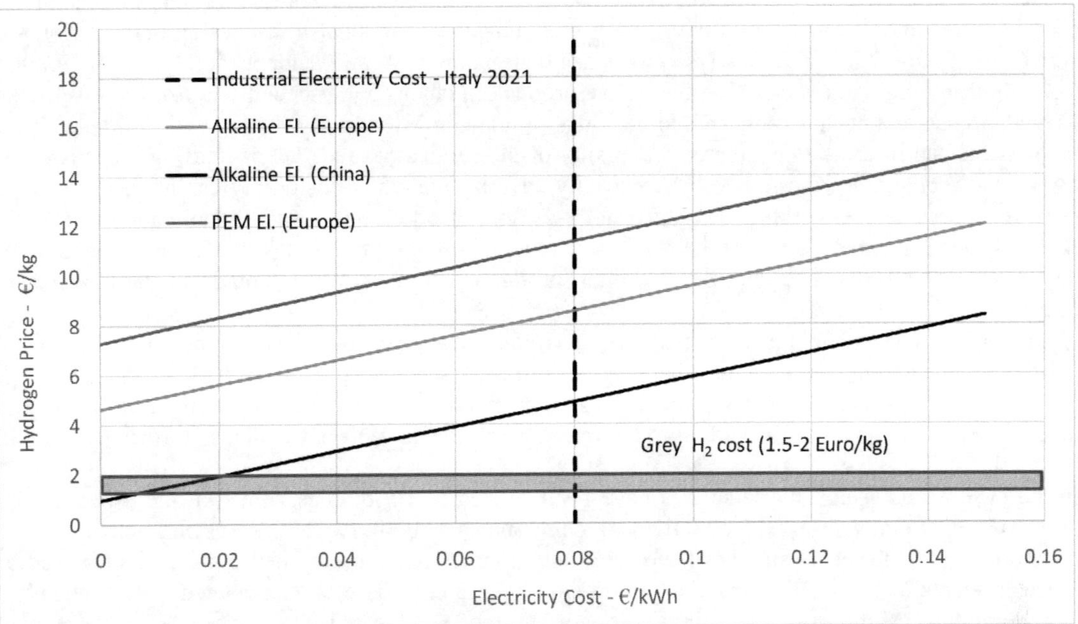

Figure 12. Overall cost of hydrogen produced by electrolysis (CAPEX included) for alkaline and polymeric electrolysers of different production.

In figure 12, with the same efficiencies used previously for alkaline and polymeric products, the straight lines are translated upwards along the ordinate axis by a value corresponding to the so-called "Levelized Cost of hydrogen" (LCOH) or the cost of producing electricity at zero price. At the current price of electricity, the cost of producing hydrogen is between 4 and 12 €/Kg, well above the current market price of hydrogen from fossil fuels (1.5-2 €/Kg) [11]. However, the production of hydrogen from electrolysis could be already convenient in the low cost area of electricity (0-0.02 €/kWh), using Asian alkaline electrolysers.
These values of the cost of electricity could be presently available using the excess energy from renewable energy (0.04 €/MWh in 2018), as shown previously. Producing hydrogen with that energy would be convenient and could be an incentive both for further investments in electric renewable energy and in the electrolytic production of hydrogen itself.
The critical aspect of this result is the cost difference between electrolysers produced in China (about 1/4) compared to those made in Western countries which allows the reduction of production costs [21]. This huge difference is explained by several factors: 1) cheaper raw materials and labor (electrolysers are still largely handmade), 2) higher factory utilization rates: China's electrolyser manufacturing industry is highly concentrated and the top three suppliers together hold a 90% share of the domestic market, 3) demand for their products (in traditional industries requiring on-site small-scale generation of pure hydrogen) stable as linked to the overall growth of the manufacturing industry, which is much stronger in China than in the West. Chinese electrolyser companies have safe and predictable sales volumes and, as a result, their production capacity is well matched to demand, resulting in high utilization rates, particularly for large electrolyser production lines [20].
Although there is a great variability in the LCOH and in its evaluation [22], it is certain that only with large electrical energyfrom renewable energy overproduction the electrolytic hydrogen can be produced at competitive cost with respect to current technologies based on cracking of hydrocarbons.

5. Conclusions

Like any change in technological paradigm, the one proposed by the Italian Ecological Transition will also require an adjustment and/or restructuring of the various industrial realities that are present

andactive in these two energy worlds. Historically, the electricity and gas sectors (and in general the electric/power and chemical industries), with their respective national transmission and distribution networks (figure 9), have always proceeded autonomously and independently, with limited technological points of contact. The respective production chains and especially transmission systems have always sought considerable autonomy and a typical top-down approach, due to the complexity of the systems and in order to guarantee the quality of the service offered. The increasing incentives for distributed production from renewable electricity have been already posing big challengesfor how to managethe reverse energy (both electricity and gas) flows (bottom-up), now and more and more in the future. This perspective highlights the need for a structural integration of the two worlds, so as to favor the national energy balance, and the hydrogen production by electrolysis could play an integrating role.
If on a traditional technological level the Italian industry appears prepared for the challenge of the Ecological Transition, [9] a sector where Italian products seem weaker today is constituted by high power electrolysis plants, which constitutes a new market space, until now limited to small and medium sizes. Competition over the production of large plants will play a decisive role in reducing production costs.
In the electrolysis sector, many studies have been carried out in the past aimed at the aspect of the efficiency and innovation of the individual components and of the entire system but always in conditions of stable power supply of the electricity source. The current need is instead addressed to other problematic aspects that are still to be explored and considered, more related to the reliability and durability of the equipment in the operating modes that will be provided for these new applications. Frequent and periodic switching on and off operations, as well as operating conditions not under constant load, will pose the need to solve problems relating to management, efficiency and duration. These problems will affect both the electrochemical section of the plants and that of the auxiliaries which, on the other hand, will have to guarantee the quality of the flows entering and leaving the plant (in particular the water and hydrogen treatment). These areas of study and experimentation, which may take place on limited but significant industrial dimensions in the demonstration plants, will have to provide useful information for the final application and the management of stacks and auxiliaries.
In conclusion, the great diffusion of electric renewable increasingly produces an excess of energy on the electricity grid and this overproduction could be used to produce of electrolytic hydrogen with already existing technologies. The priority use of this "green" hydrogen could cover part of the current national industrial demand, thus fulfilling the dual function of reducing carbon dioxide emissions from conventional production processes and providing a method to simplify the management of electric renewables in overload conditions. In parallel or nextly, electricity / gas integration could perform the same function by allowing hydrogen (P2G) to be fed into the network.
The analysis carried out shows that a few large electrolysis plants located where there is and will be an increasing surplus of renewable electricity (mainly in Southern Italy) with many modular electrolysis units could fulfill this function. Inexpensive alkaline electrolysers seem to be preferred for their reliability and durability and for the production of hydrogen at competitive costs.

References

[1] ARERA, *Relazione sullo stato dei servizi*, 321/2020/I/FER - https://www.arera.it/allegati/docs/20/321-20.pdf
[2] Ministero per lo sviluppo Economico, *Piano Nazionale Integrato per l'Energia e il Clima*, https://www.mise.gov.it/images/stories/documenti/PNIEC_finale_17012020.pdf
[3] ARERA, *Relazione sullo stato dei servizi*, 321/2020/I/FER - https://www.arera.it/allegati/docs/20/321-20.pdf
[4] Terna S.P.A., *Contesto ed evoluzione del sistema elettrico*, 2019, https://download.terna.it/terna/Contesto%20ed%20evoluzione%20del%20Sistema%20Elettrico_8d75639fa148d01.pdf

[5] RSE, *Energia elettrica, anatomia dei costi*, 2018, http://www.rse-web.it/applications/webwork/site_rse/local/doc-rse/Energia_Elettrica_Anatomia_Costi_2019/index.html
[6] GSE, *Rapporto delle attività 2020*, https://www.gse.it/servizi-per-te/news/online-il-rapporto-attivita-2020
[7] Jopek A.G., *Hydrogen production by electrolysis*, Wiley, Monaco, 2015
[8] Winter C.J., *Hydrogen as an Energy Carrier: Technologies, Systems, Economy*, Springer-Verlag, Berlino, 1988
[9] Terna, *Progetti pilota di accumulo*, https://www.terna.it/it/sistema-elettrico/innovazione-sistema/progetti-pilota-accumulo
[10] 2011 Fuel Cell Bullettin**7** 9.
[11] Terna, *Dati sulla produzione di energia elettrica in Italia 2020*, https://www.terna.it/it/sistema-elettrico/statistiche/pubblicazioni-statistiche
[12] SNAM, *The Hydrogen challenge: the potential of hydrogen in Italy*, https://www.snam.it/it/hydrogen_challenge/repository_hy/file/The-H2-challenge-Position-Paper.pdf
[13] https://industria.airliquide.it/inaugurazione-del-piu-grande-elettrolizzatore-pem-al-mondo
[14] https://www.asahi-kasei.co.jp/asahi/en/news/2018/e180810.html
[15] https://www.refhyne.eu/
[16] Fuel Cells and Hydrogen Observatory 2020, *Hydrogen molecule market*, https://fchobservatory.eu/sites/default/files/reports/Chapter_2_Hydrogen_Molecule_Market_070920.pdf
[17] www.nelhydrogen.com
[18] International Renewable Energy Agency, *Green hydrogen cost reduction*, 2020, https://irena.org/-/media/Files/IRENA/Agency/Publication/2020/Dec/IRENA_Green_hydrogen_cost_2020.pdf
[19] Levene J.I., Mann M.K., Margolis R.M., Milbrandt A., 2007 *Solar Energy***81**773.
[20] Eurostat Statistics, 2020, https://ec.europa.eu/eurostat/statistics-explained/index.php?title=Electricity_price_statistics
[21] Bloomberg New Energy Finance. *Hydrogen: The Economics of Production From Renewables*. Tech. rep. 2019.
[22] Pozio A., Bozza F., Nigliaccio G.,Platter M., Monteleone G., 2021 *Energia Ambiente ed Innovazione***1**66
[23] Christensen A., *Assessment of Hydrogen Production Costs from Electrolysis: United States and Europe*, International Council on Clean Transportation, 2020, https://theicct.org/sites/default/files/publications/final_icct2020_assessment_of%20_hydrogen_production_costs%20v2.pdf

Innovative PV Technologies for reducing electricity costs

Cosimo Gerardi*[1], Giuseppe Bengasi[1], Lorenzo Carbone[1], Antonio Spampinato[1], F rancesco Rametta[1], Antonino Ragonesi[1], Gaetano Izzo[1], Marcello Sciuto[1], Marina Foti[1], Fabrizio Bizzarri[2]

[1] 3SUN, ENEL Green Power, Contrada Blocco Torrazze 95121, Catania, Italia
[2] ENEL Green Power, Viale Regina Margherita 137 00198, Roma, Italia

* Corresponding Author: Cosimo.Gerardi@enel.com

Abstract. Solar cells cost per Watt has decreased so quickly in the last years that photovoltaic (PV) module manufacturing costs cover only a small fraction of the total costs of a PV plant. Essentially, the installation and operation costs as well as the costs for the management of starting materials are the main factors today the most important levers for total costs decrease. PV technology has advanced towards both efficiency and average energy generation increasing by light capture optimization and losses minimization. This work analyses the state-of-the-art of PV technology for terrestrial application and discusses the perspectives for future development, which aims at overwhelming limitation posed by material properties. Higher focus is given to the most promising current technology such as silicon heterojunction solar cells and on future developments based on tandem cell approaches.

Keywords. Photovoltaic System, Cell Efficiency, Module, LCOE, PV Market Trend

1. Introduction

In the last years, photovoltaics systems have shown to play a pivotal role in the transition toward renewable energy sources and CO2 emissions reduction. PV sustainability has remarkably improved due to technology progresses, which have led to a lower materials waste and an increased manufacturing yield. Moreover, improving the device architecture resulted in higher energy conversion efficiency and in higher overall annual electricity generation.

Herein we report an overview of PV technology for terrestrial applications, mainly based on crystalline silicon. The mainstream technologies are based on device architectures and processes proposed in the 80's that have been optimized along the last decades to achieve maximum performance and lower costs.[1,2] Indeed, recent solutions, such as bifacial modules, highly improved the PV performances in terms of energy production.[3–5] Chinese manufacturers dominate the terrestrial PV market achieving more than 90% of the PV worldwide production, leading the technological breakthrough required from the global needs. In this work, we report the advantages of an innovative silicon-based technology, known as heterojunction technology (HJT). Importantly, we disclose how the technological advantages of HJT solar cells could place the PV European manufacturers in a leading position.

Finally, we will discuss the future technology evolution based on new architectures (tandem) and materials (perovskite), which will allow solar cells to significantly increase the conversion efficiency.

2. The PV Market

Crystalline silicon technology dominates the world PV scenario owning 90% of the market shares as shown in Figure 1 (right). This technology relies on either multicrystalline or monocrystalline substrates for the solar cell manufacturing. Essentially, the monocrystalline silicon technology role in the PV market is nowadays raising up due to its higher efficiency with respect to the multicrystalline systems, compensating the higher manufacturing costs with higher energy delivery.[8] The remaining 10% of the market share is employing the so-called thin film technology. Thin films solar cells are produced depositing cadmium telluride (CdTe) and CuInxGa1-xSe2 (CIGS) thin films on glass substrates.[9] This kind of solar panels are becoming part of a niche, mainly focusing on alternative markets such as building integrated photovoltaics system.[10]

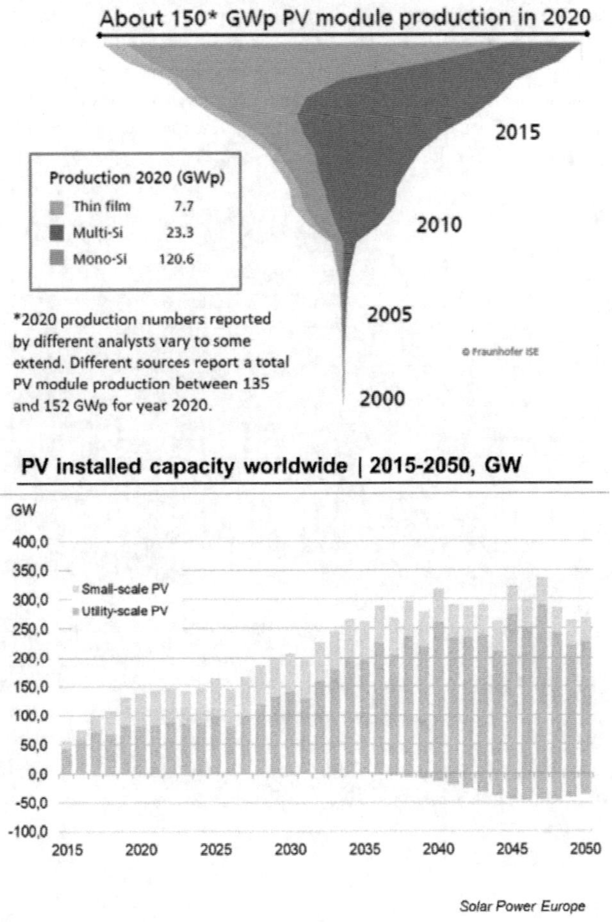

Figure 1: Up: Terrestrial PV technologies market shares (source: Fraunhofer ISE 2021[6], which estimates that in 2020 about 150 GWp were manufactured). Bottom: PV installations per year (data Solar Power Europe 2020 [7]). The negative values refer to PV modules which are expected to be replaced in the next years.

If compared with crystalline silicon PV, thin film modules are limited in terms of conversion efficiency. Moreover, some thin film technologies make use of rare or toxic materials and in general they require specific resources for assembling the modules, reflecting an increase of the billing of material costs (BOM) which hindered the scientific efforts to further improve the system.[11,12]

In the last years, crystalline PV modules have overcome efficiencies of 19% on areas of about two square meters. The most performing modules achieve even more than 21% and are mostly commercialized for "high-

end" residential market at higher costs.[11] The cumulated global capacity of installed modules has passed the 500GWp threshold and it is forecast to reach more than 1 TWp by 2023. [13] The Watt peak (Wp) represents the measure of the maximum power output of a PV module, or viceversa the maximum power that can be absorbed by an electric load. Moreover, the annual PV modules shipment overcomes 100 GWp and it is forecast to grow up in the next years favoured by the incredible price reduction observed in the last period. [8] The levelized cost of the energy (LCOE) is a reference value for the price of the energy per generated electric energy unit, in order to have an adequate economic return of the financing and maintenance costs of the solar plant along its life cycle.
The PV market will be further conditioned by a crucial parameter, not only based on the module costs, but also related to the module performances and reliability considering the decrease of the final energy cost, which evolves in the levelized cost of energy factor. The LCOE is expressed in €/kWh and it used as an estimation of the economical technology convenience. The LCOE is the ratio between the average cost required to finance and maintain the energy plant along its useful lifecycle, and the generated total amount of energy in the same period. [14] The increased demand of PV modules and the technology advancement is constantly leading to lower PV module costs as reported in Figure 2.

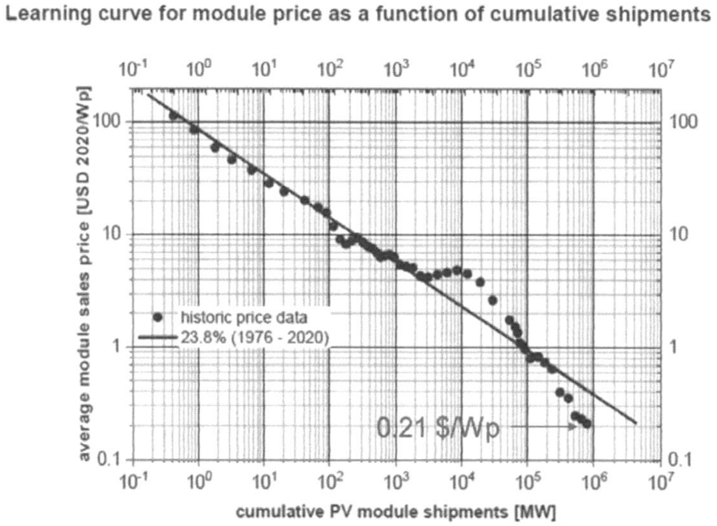

Figure 2: Average PV module sales prices as a function of the cumulative shipments (source: International Technology Roadmap of Photovoltaics 2021).[15]

3. Perspective for a future with more Photovoltaics

The photovoltaics technology is an important player in the race for global warming control. Indeed, PV technology provides one of the most environmentally friendly solutions for energy production triggering scientific efforts to increase the system efficiency aiming to reduce the production costs. Indeed, recent research results shed light on outstanding perspectives to further increase solar cells efficiency, leading to cost reduction levels which were unconceivable few years ago. These module cost reductions have contributed to an even more significant reduction in the levelized cost of photovoltaic generated electricity.
The photovoltaic module efficiency is a major driver for the PV industry. This impulse arises because cell processing costs are subject to the same downward price pressure as in microelectronics, through increased manufacturing volume and sophistication processing. However, other costs, such as those of glass, polymer encapsulant, aluminum frame, junction box and cell interconnectors do not benefit to the same extent. The cell efficiency improvement directly decreases all these less controllable costs reflecting the reduced amount required for a given energy output.
Moreover, the cell efficiency increase will likely lead to reduction in PV modules area and weight, implying the opening towards new markets development such as building integrated PV and transportation sector.
Despite the high enthusiasm for PV technology progresses and the interesting development perspectives, European PV market is experiencing a significant crisis. On the contrary, Chinese industry is very healthy and dominates the market with more than 90% of share. Nevertheless, Europe can still rely on robust pillars given by technology competence and industrial know-how, and it can reconquer its market share to become a primary player in the energy transition. Through technology innovation, there is the opportunity to rebuild a leading PV industry in Europe, able to compete with mainstream technology.

3.1 Beyond the Cell Limits...

The I-V curve of an ideal solar cell exposed to the light can be calculated from the balance that involves electrons and photons at each voltage value. This allows to obtain the theoretical limits of solar cell efficiency. The limit can be derived from the Shockley-Queisser (S-Q) [16] theory in which an ideal cell is in the dark and at the thermal equilibrium so that it would absorb all the incident radiations given in this case by the blackbody radiation at room temperature. Through the detailed balance principle, the cell must also emit that radiation (see a schematic effect view on the left side of Figure 3). Since the cell is in dark conditions, the source of the photons emitted with energy higher than Eg is given by the transition to the fundamental state of the charges, which are in the excited state. Many photons emitted in this way are reabsorbed by the cell before being extracted as current between the electrodes. Integrating the number of photons in the black body radiation emitted in the energy interval above Eg, the total radiative recombination rate in the device at the thermal equilibrium can be obtained. Some current PV technologies are approaching the radiative limits which are related to the absorber Eg in the graph shown in the right side of Figure 3. Table 1 reports the record efficiencies obtained in laboratory for several types of materials.

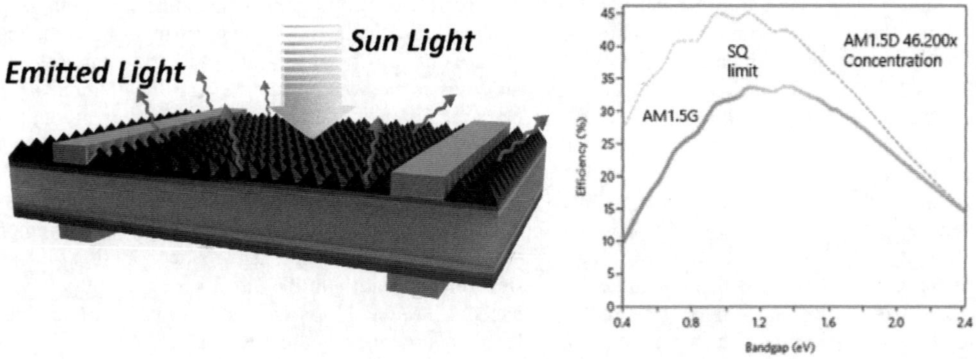

Figure 3: (Left) Scheme of a HJT solar cell. (Right) Solar cell efficiency theoretical limits. Shockley-Queisser limits for both AM1.5G and concentrated (46200x) spectral conditions.[17] AM1.5G is the air mass coefficient, which defines the length of the light path across the Earth atmosphere. It is expressed as the ratio

related to the optical path at a given angle defined with respect to the zenith. AM1.5 refers to the value corresponding the path at about 42° with respect to the zenith, and it is taken as global reference for terrestrial PV (G: global). For silicon (Eg = 1.15 eV) the theoretical limit at AM1.5G is of 29.5%.

Material	Organization	h (%)
Si	Kaneka (Japan)	26.7
μc-Si:H	AIST (Japan)	11.9
a-Si:H	AIST (Japan)	10.2
GaAs	Alta (China-USA)	29.10
InP	NREL (USA)	24.20
GaInP	NREL (USA)	22.00
CIGS	Solar Frontier (Japan)	23.35
CdTe	First Solar (USA)	21. 00
CZTSSe	DGIST (S. Korea)	11.30
CZTS	UNSW (Australia)	10.00
Perovskite	UNIST (S. Korea)	25.50[18]
OPV	SJTU (China)	18.07
Dye	Sharp (Japan)	11.90

Table 1 Best laboratory solar cell efficiency record. h is the conversion efficiency. In case of silicon the record value is obtained on cell with sizes similar to industrial solar cells. In the other cases the sizes are quite lower and, in some cases, (as Perovskite, CZTS, CdTe, GaInP) the cells are small and not at industrial scale.[19] Eth is the threshold energy that a photon must have in order to be absorbed by the material and it is given by the band-gap minus the exciton energy level.

3.2 PV Technologies for Terrestrial Application

PV technologies for terrestrial use can be divided in two families: i) Crystalline silicon PV and ii) Thin films PV. To the first family belong the most widespread technologies that are based on assembling solar cells obtained by processing polycrystalline or monocrystalline silicon wafers. In particular, the use of monocrystalline silicon has grown in the last years allowing PV modules to significantly increase their performances.

Monocrystalline silicon wafers are obtained by using elevated temperature processes and are produced from ingots grown from molten silicon. The most diffused technology employed for realizing the ingots is known as Czochralski, or CZ method.[20] The method employs a crystal seed which is put on a rod and dipped in molten silicon obtained by melting polycrystalline silicon at temperatures of 1420°C [21]. The seed rod is slowly pulled upwards and simultaneously rotated. By precisely controlling the temperature gradients, rate of pulling and speed of rotation, it is possible to extract a large, single-crystal, cylindrical ingot from the melt. The silicon wafers, with thicknesses of 150 - 200 μm, are obtained by sawing the cylindrical ingot that has been transformed in a parallelepiped shaped brick. The cut is carried-out by using a very thin diamond wire.

Once processed to form a diode with the electrodes for extracting the current, the solar cells are interconnected in a photovoltaic module. Typically, a solar cell has size of 166 mm ´ 166 mm, and a PV module is made by connecting 60 or 72 cells. The cells are connected in strings as shown in the upper part of Figure 4.

Today, due to the remarkable improvement of crystalline silicon quality it is easy to obtain PV modules with efficiencies between 18% and 20%. For high-end, higher cost, products the efficiency can overcome 21%.[19,22]. Moreover, to enhance the electricity production some of the crystalline silicon modules existing on the market are bifacial. Bifacial modules are designed to capture both the direct sunlight and part of the light reflected or diffused from the ground. The reflected or scattered light is collected from the rear side of the solar cells, contributing to increase the average electricity produced by the modules.[23] More information is discussed in the following chapter.

The second group of PV technologies includes thin film modules that are obtained depositing few microns of absorber materials on large area and low-cost substrates such as glass, metal, or plastic. One of the main challenges related to thin film technologies, is to maintain the uniformity on the large substrate area. In fact, on large area it is particularly complex to achieve homogeneous thickness, chemical structure and physical properties. Therefore, thin film solar modules struggle to compete with crystalline silicon technology.[24,25]

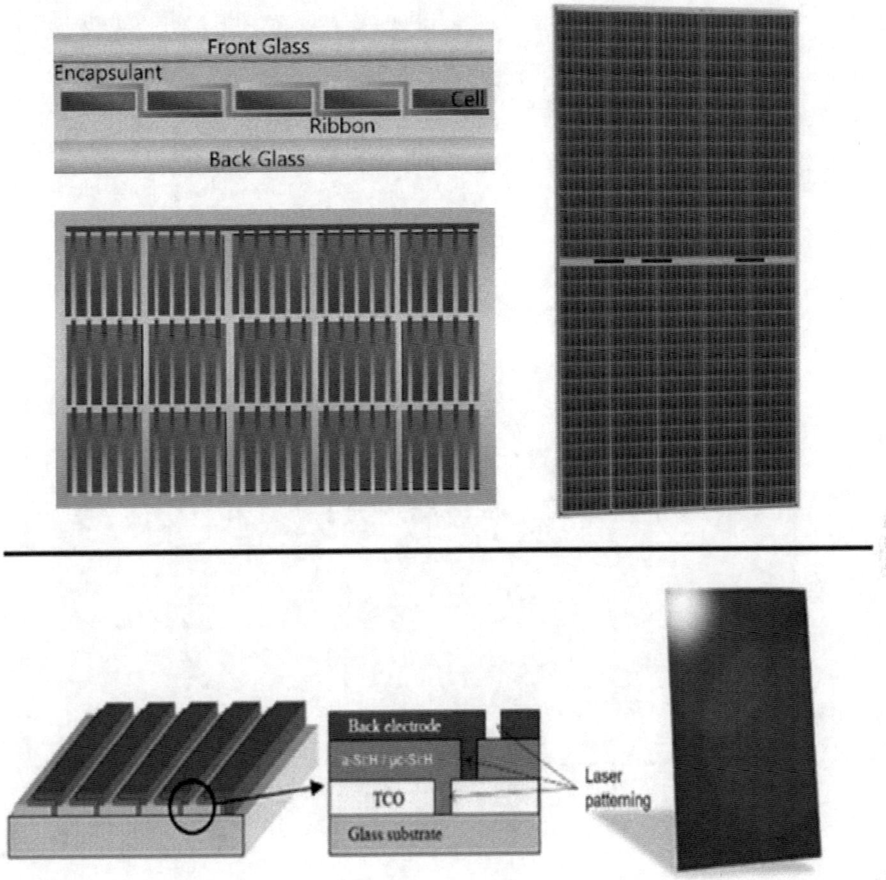

Figure 4: Comparison of crystalline silicon technology (top) based on the assembly of solar cells thin film technology (bottom) (considering the case of amorphous silicon).

Typically, thin film modules use a monolithic approach, consisting in sequential material deposition and laser ablation to define active areas, interconnections and insulating regions schematically reported in the bottom part of Figure 4.[26] Another key aspect in thin film technology is the design of light trapping systems aiming to increase the light path and, consequently, increase the number of absorbed photons. Generally, light trapping is achieved incorporating sub-micrometer structures in the glass or on the transparent conducting oxides (usually aluminum or boron doped zinc oxide or fluorine doped tin oxides). [27–29] Owing to these technologies, inspired by the structure allowing night vision in insects' eyes (e.g moths), it is possible to significantly increase the equivalent thickness of active material. The category of thin film includes materials such as cadmium telluride (CdTe) or Copper Indium Gallium Diselenide (CuInxGa1-xSe2) known as CIGS (see table 1). Other materials such as amorphous silicon and microcrystalline silicon as well as organic materials are characterized by lower efficiency and are mainly used for particular applications such as building integrated PV due to their process easiness and low costs. Recently, thin films solar cells based on hybrid organic – inorganic perovskite absorbers attracted a lot of attention for the possibility to realize high efficiency solar cells at very low cost as following described in the next paragraph.[18,30]

Thin film PV experienced a growing period between 2008 and 2012. Such growth was related to the crystalline silicon supply crisis. However, the crisis has been temporary and crystalline silicon technology has returned to dominate the market as reported in Figure 1 and Figure 5. The latter figure reveals the market shares trends of the main technologies that are based on the usage of silicon, both crystalline and polycrystalline. In the meanwhile, crystalline silicon imposed itself in the market thanks to the standardization of the supply chain of materials such as silicon wafers, glasses, polymeric encapsulants, interconnecting materials, junction boxes and cables. Because of these advantages, crystalline silicon modules are characterized by lower process costs both in terms of cost/W and of cost/kWh, due to superior reliability and higher average electricity supply.

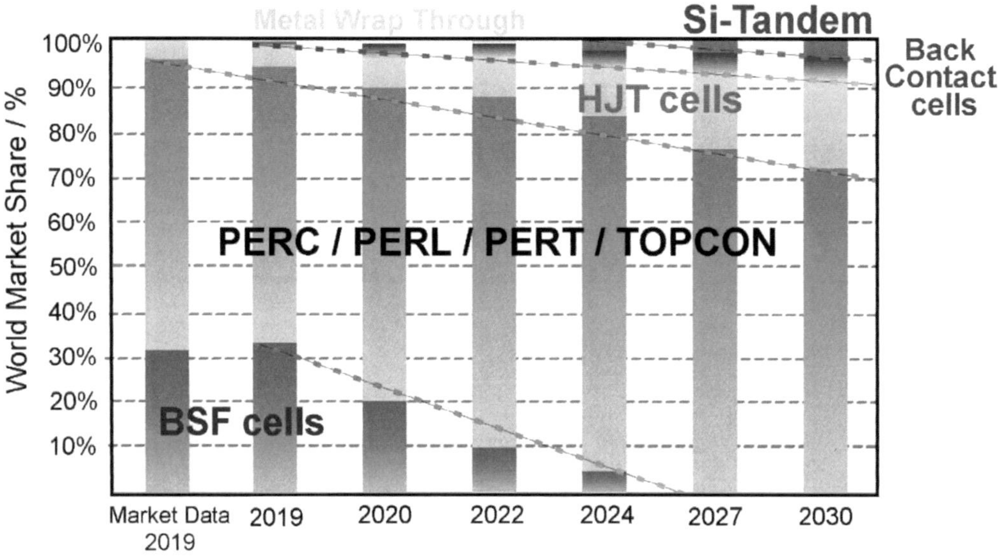

Figure 5: Market shares trends of the main technologies that are based on the usage of silicon [9] (from ITRPV 2020)

4. Crystalline Silicon Technologies

4.1 Al-BSF technology

Up to few years ago, the most diffused crystalline solar cell technology was known as aluminum back surface field (Al-BSF), which utilizes a continuous layer of Aluminum as backside contact (Figure 6).

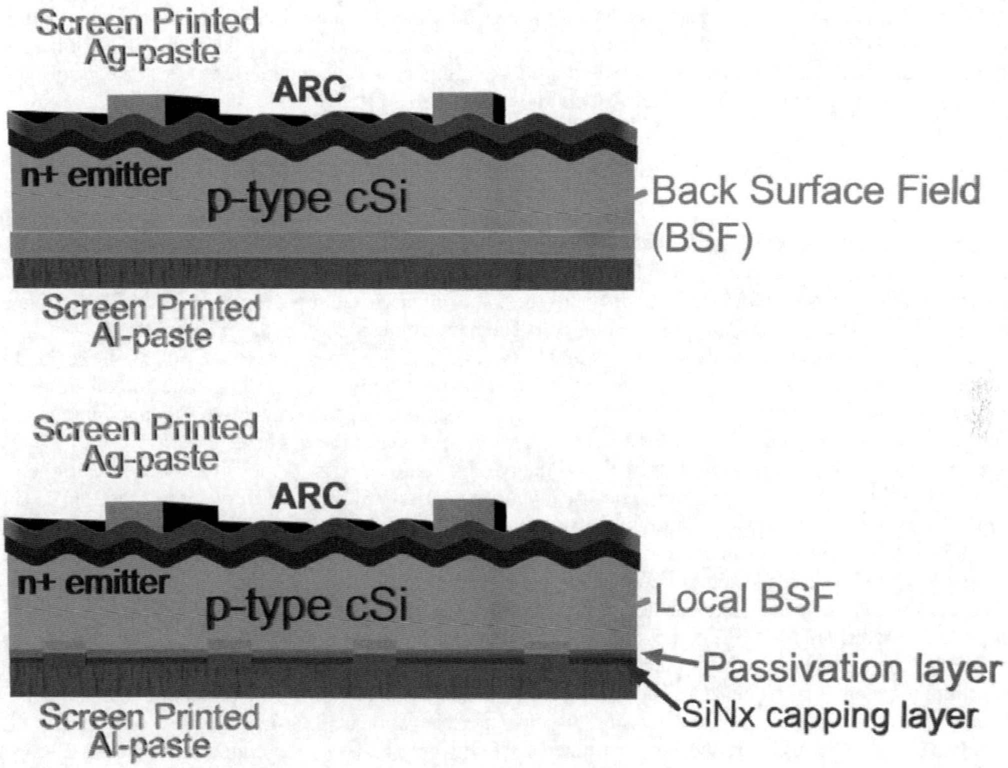

Figure 6: (top) Al-BSF solar cell structure. In the front side of the cell there are the narrow metal stripes (fingers), with a width of about 40-60 μm, obtained with screen printing. In the rear part of the cell there is the continuous layer of aluminum. The high temperature firing process forms an aluminum silicon alloy, which gives rise to the back surface field. (bottom) PERC cell architecture. One can note in the rear side of the cell the local openings made in the passivation layer, which consists of a stack of Al_2O_3 and Si_3N_4. In the contact opening regions, the BSF is formed.

Typically, Al-BSF technology (Figure 6) employs p-type multicrystalline or polycrystalline silicon, doped with boron. The n-p junction is formed by a high temperature diffusion process, by which the n-type dopant, typically phosphorus, penetrates into the p-type silicon, giving rise to the n-p junction at the front of the solar cell. Only the front surface of the cell is passivated by depositing a dielectric, typically silicon nitride, which acts also as an antireflection coating (ARC). The Al-BSF process workflow can be resumed as following: after an initial inspection, the silicon wafers are treated in a chemical wet bench using solutions based on KOH. Due to the different reaction speed of the etching process on the (100) crystal plane compared to the (111) direction, the KOH treatment leads to the formation of pyramids with heights of few microns. The pyramids have the function of reducing the reflection and distributing the light obliquely in the silicon. As a result, the cell optical absorption increases. After the chemical treatment, the wafers are subject to a high temperature diffusion process, using POCl3 as precursor of phosphorous, which diffuses in the p-type wafer. The cell contacts are obtained printing metallic pastes as a grid on the front side and as continuous metal layers on the back side. Screen printing processes represent a key technology for the crystalline solar cell manufacturing. It is used to print either aluminum in low-cost processes or silver in high performing devices. In Al-BSF cells, after printing the contacts, the remaining organic solvents present in the metal pastes are stripped away through drying processes. Finally, a firing process at about 800°C is needed for metal sintering and to achieve an alloy between the metal and the silicon. The backside contact builds up an electric field at the backside interface of the solar cell (back-surface field, BSF). This electric field has a very important function in the device since it drastically reduces the surface recombination velocity of charge carriers.[31,32]

4.2 PERC Technology

In the last years, most of PV manufacturers have switched their production from multicrystalline to monocrystalline silicon (see Figure 1), due to the overall reduction of costs/Watt for monocrystalline silicon solar cells manufacturing. To further increase the efficiency and obtain more benefits from the use of monocrystalline Silicon, the device architecture of the solar cell has been improved by introducing the concept of passivated emitter rear contact (PERC) solar cell. The new architecture allows a more effective passivation of the surface thus drastically reducing the recombination velocity. In particular, PERC solar cell architecture introduces the passivation of the n-doped layer, known as emitter.[33] With PERC architecture, a significant improvement in terms of efficiency has been achieved in crystalline silicon solar cells. Although more process steps are necessary to fabricate the PERC cells compared to Al-BSF, the increase in performances allows a significant lowering of costs/Watt. In PERC cell the passivation of the front side is extended also to the rear side of the cell, by depositing a stack of dielectric layers on the back side of the wafer (Figure 6 bottom). PERC solar cells feature rear passivation through an AlOx/SiNx dielectric stack, which has an effective rear surface recombination velocity of about 100 cm. Light management is also improved by optimizing layer thicknesses of the AlOx/SiNx stack, and therefore excellent rear internal reflectance is achieved. In Al-BSF the direct contact of the continuous metallic contact introduces many recombination centers. In PERC the rear contact is obtained through small openings in the dielectrics achieved through an infrared laser (local contact openings). In this way the electronic quality of the contact at the interface between metal and silicon is improved, thus reducing the recombination centers. The firing process leads to the sintering of the metallic contact, and the thermal treatment defines the formation of the BSF regions only in the small local contact areas. The recombination can be further reduced by defining locally diffused regions with high concentrations of dopant (typically phosphorous in p-type standard PERC) just below the contact with the metal.[34] Despite some improvement in rear surface passivation, the loss analysis studies still suggest that recombination at the rear surface (especially at the rear metal–Si interface) is the efficiency limiting factor of the PERC architecture.[35]

4.3 New Module Designs

The architecture of the solar cells has been almost the same since the 80's. Since then, using the same device architectures there has been a significant development in the materials performances as in the case of the silver pastes that have led to a remarkable improvement of the grids with better aspect ratios, very thin fingers and optimized antireflection coatings.[36–41] It has taken about 30 years to obtain commercial solar cells with the efficiency levels originally obtained in the research laboratories. In the last few years, there have been revolutionary changes in the architecture of the crystalline silicon modules thanks to the introduction of more sophisticated assembly schemes. These new schemes reduce cell to module (CTM) losses related to dead areas and resistive losses. The new architectures also take advantage of bifacial solar cell designs, which will be described in the following section. To improve solar panel active area, some manufacturers have introduced the concepts of half cut cell assembly and of shingling.

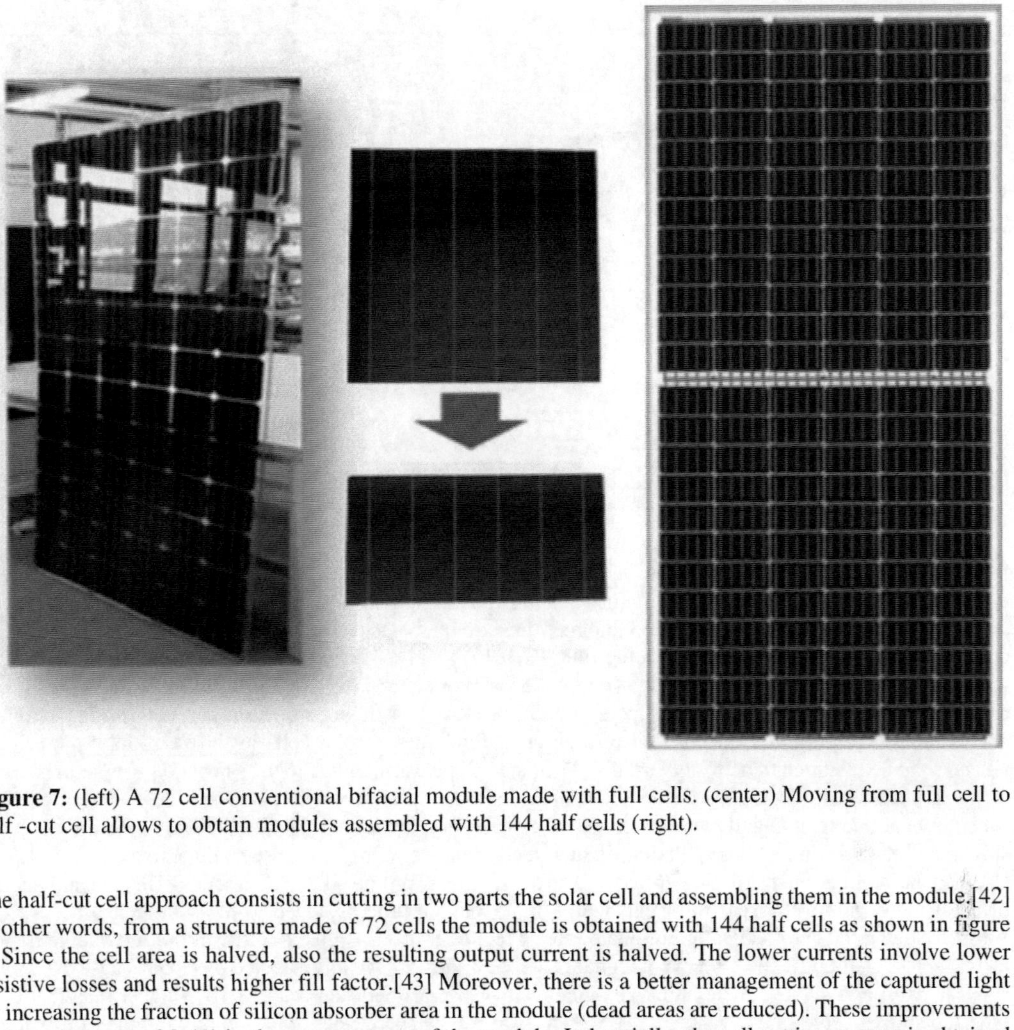

Figure 7: (left) A 72 cell conventional bifacial module made with full cells. (center) Moving from full cell to half -cut cell allows to obtain modules assembled with 144 half cells (right).

The half-cut cell approach consists in cutting in two parts the solar cell and assembling them in the module.[42] In other words, from a structure made of 72 cells the module is obtained with 144 half cells as shown in figure 7. Since the cell area is halved, also the resulting output current is halved. The lower currents involve lower resistive losses and results higher fill factor.[43] Moreover, there is a better management of the captured light by increasing the fraction of silicon absorber area in the module (dead areas are reduced). These improvements yield an increase of 3-4% in the output power of the module. Industrially, the cell cutting process is obtained by laser cut and mechanical cleaving of the two half pieces. The cut process is of crucial importance. In fact, the cut can induce damages in the cell and deteriorate both electronic properties and mechanical robustness of the solar cell.[44] As consequence, a not-appropriate cut may lead to lower efficiency and/or reliability issues.[45,46] Furthermore, the new surface created by the cut, suffers from a poor passivation[47] and may require changes in the other processes such as $Al2O3$ deposition or light soaking to improve efficiencies.[48] Another strategy, named shingling, consists in the cut of the cells in smaller parts and connecting these shingles employing Electrically Conductive Adhesive (ECA).[49] (Figure 8). Owing to this approach, shingling technology could enhance the efficiency of 8-10% compared to classic PV modules due to lower resistive losses and increased surface area. [50]

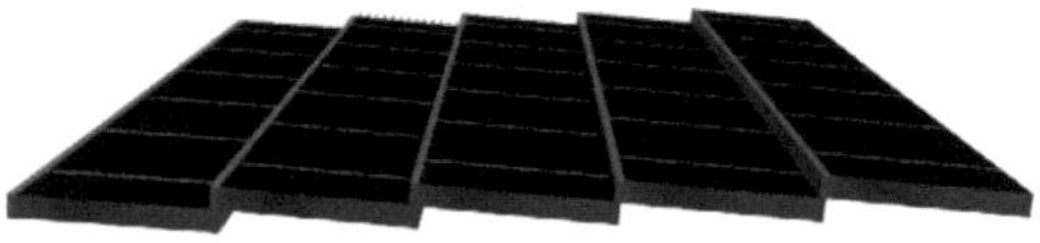

Figure 8. (left) Scheme of a shingling assembly. (right) Two shingled modules, obtained using HJT solar cells fabricated in the 3SUN EGP Factory, during an outdoor testing.

In general, reliability of PV modules is very important to establish their lifetime in a PV solar plant. Although the overall improvement of materials quality and manufacturing processes, reliability problems can represent an important issue in modern PV technologies. In particular, PV modules suffer from several reliability problems arising from the exposure to the light itself [51] and to environment agents, such a moisture,[52] mechanical stress (wind and snow),[53,54] migration of mobile ions from the glass into the silicon,[55] etc.
Crystalline silicon PV modules are affected by deleterious light induced degradation (LID). [51] The LID phenomenon was observed for the first time in crystalline silicon solar cells in the 70's.[56,57] When the silicon solar cells started to be widely used, LID created some warning due to the impact on long term stability of solar plants. LID degradation is related to the boron-oxygen complexes,[58,59] defects formed by a substitutional boron atom and two oxygen atoms. These complexes are typically present in the p-type doped silicon and the creation and annihilation of such metastable complexes are thermally activated, though they also depend on light intensity. Therefore, LID strongly affects p-type silicon solar cells. This phenomenon is faced at industrial level and is mitigated with a light soaking process added to the solar cell process flow. Another strategy able to reduce or eliminate effects related to LID, is linked to the use of alternative doping agent such as Gallium[60,61]. On the other hand, n-type wafers are weakly affected by LID because the doping agent is phosphorus. Nevertheless, n-type silicon wafer price is between 5% and 10% higher than that of p-type. This aspect gives the advantage of lower costs to p-type silicon-based PV, improving their diffusion in the market with respect to n-type.
On the other hand, monocrystalline silicon ingot fabrication technology has evolved in terms of costs reduction, making even more convenient, in terms of cost/Watt, the employment of high performing monocrystalline silicon. This is leading to an increasing use of n-type wafers because of their better performances. In fact, the use of n-type monocrystalline silicon brings several advantages such as lower intrinsic recombination and lower Shockley-Read-Hall (SRH) recombination caused by contaminants such as iron.[62] Hence, the average minority carrier lifetime (MCL) is remarkably higher in n-type silicon [63]. Besides higher performances, n-type silicon based solar cells are characterized by a better reliability and lower degradation owing to reduced LID effects related to the absence of boron.
More recently a new phenomenon of degradation in p-type silicon has been observed induced by light and high temperatures. This phenomenon is known as Light and elevated Temperature Induced Degradation (LeTID) and can lead to significant power reduction (even more than 5%).[64]
Another important degradation phenomenon is known as Potential Induced Degradation (PID) and it is related to the migration of sodium ions or other mobile impurities from the glass into the silicon, causing current losses that lead to a drastic degradation of the PV module.[55,65,66]
The PID degradation can be reduced or controlled with barrier layers, which block the migration of the impurities. For this reason, the polymeric encapsulant plays a key role acting as barrier layer against both

sodium and moisture.[67] The most diffused encapsulant are based on ethyl vinyl acetate (EVA) or on polyolefin (POE), which are laminated with the front glass and the module backsheet.

4.4 Bifacial Modules Technology

The bifacial modules technology has recently consolidated on the market. Bifacial modules use both front and rear faces of the solar panel to capture the sunlight.[23] While the front side will capture the direct light impinging on the cell, the backside face captures the light reflected or scattered from the ground (Figure 9). The bifacial option can offer an average increase of energy conversion from 5% to 20% and even more.[3,5,68] The conversion increase is highly dependent on the reflectivity of the ground or albedo. Bifacial modules market share has remarkably grown in few years and a continuous growth is expected in the next years. In fact, it is estimated that bifacial modules market share will increase up to 30% and even more in the next years.[8] At module level, the main difference compared to standard monofacial technology, is the architecture of the module that must be changed replacing the opaque Tedlar backsheet with another glass or a transparent backsheet. The bifacial solar module also requires a different solar cell architecture. In particular, the continuous metal back contact is not compatible with the bifacial approach, and it must be replaced with a metallic grid.[23] This architecture allows light to enter in the absorber layers also from the cell backside resulting in a bifacial solar cell. However, replacing the continuous metallic back contact with a grid reduces the contact area increasing the contact resistance and decreasing the fill factor.

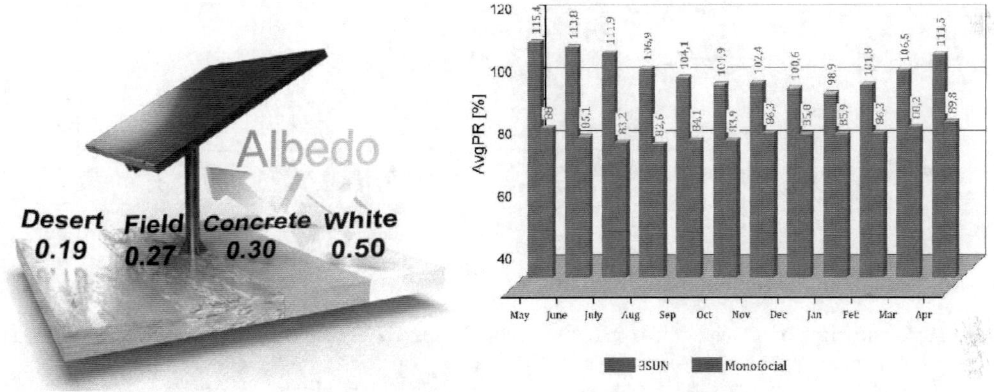

Figure 9:(left). Scheme representation of bifacial modules operation and different albedo values related to different types of grounds. (right) Comparison of performance ratio between monofacial PERC modules and HJT bifacial module fabricated by 3SUN factory of Enel. The high bifacial performances of HJT PV modules allows a significant higher performance ratio compared to monofacial PERC technology.

Hence, a trade-off must be found between the amount of light that passes across the grid (grid shadowing) and preserving the fill factor of the cell to avoid the contact resistance increase. The trade-off determines the final performances of the bifacial solar cells.[69] The performances of bifacial solar cells are evaluated on two parameters: i) the efficiency measured from the front ii) the bifacial factor given by the ratio between the efficiency measured from the backside and that measured from the front. To achieve the best performances the ideal solar cell must be symmetric to have the highest bifacial factor. Achieving good symmetry in the solar cell is not trivial and it is limited by the solar cell architecture. Actually, PERC solar cells are not symmetric as the backside of the cell is less transparent. The bifacial variant of PERC solar cell is known as PERC+ and it requires a further optimization of the passivation layer. Typically, PERC+ solar cells exhibit a bifacial ratio of 65% to 75%.[70] Exceeding such values is particularly difficult because of the thick passivation layer on the backside of the PERC cell. Other architectures are more symmetric and can give more output with bifacial factors exceeding 90%.[71] One of the best options is represented by silicon heterojunction technology, which is very symmetric and will be discussed in the following section. Figure 10 displays a comparison of bifacial ratio of different technologies, essentially PERC+, PERT and HJT, as measured at the flasher from the front and backside of the modules. PERC+ modules from different suppliers exhibit a bifacial factor ranging below 75%. PERT solar cells are more symmetric than PERC and exhibit a higher bifacial characteristic. HJT modules show the best bifacial factors, in that case cells with bifacial factor higher than 95% have been assembled in modules by using different materials. The best results are achieved by the optimized HJT module structure.

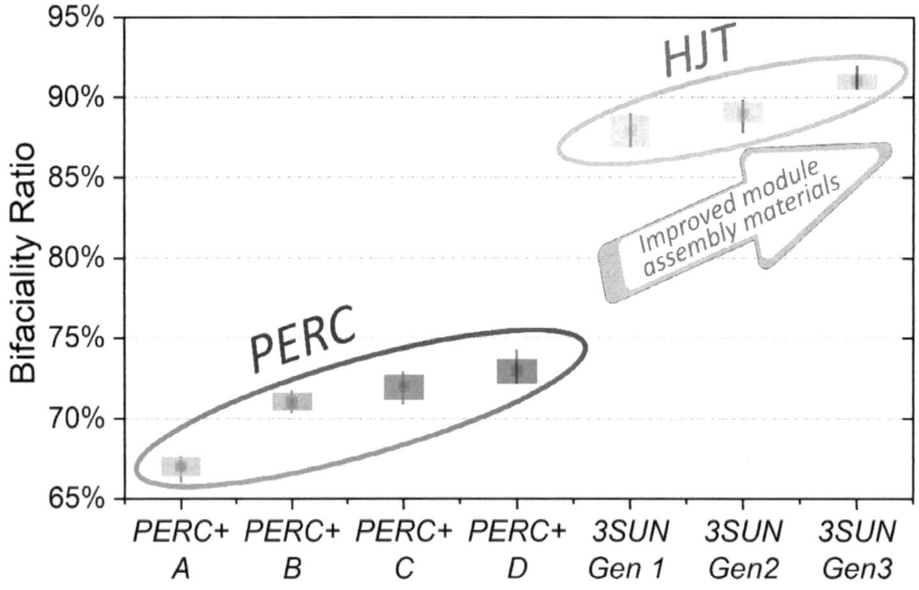

Figure 10: Comparison of bifacial ratio of different technologies. HJT exhibits the best bifacial factor. Module materials optimization leads to best performances for bifacial properties.

4.5 Silicon Heterojunction Technology

Improving the efficiency of solar cells is crucial to favor the utilization of the photovoltaic energy. Essential device properties, such as minority carrier lifetime, series and parallel resistances as well as optical properties must be all optimized to reduce recombination losses and optical and resistance losses.[72] The progress of monocrystalline silicon electronic properties has been a very important lever to increase the efficiency of solar cells. Fundamentally, solar cells must generate the biggest number of charge carriers by absorbing the light. The charges must survive as much as possible without recombining until they reach the electrodes and are collected giving rise to the solar cell current. This behavior is central to avoid the recombination at the silicon wafer surfaces, which becomes very important as the thickness of silicon wafer is reduced. In the years, several approaches have been introduced to reduce surface recombination.[73–75] Nevertheless, though the dielectric used for passivation causes a significant surface recombination velocity decrease, the direct metallic contact with the silicon introduces a large amount of recombination centers in the contact region, thus limiting crystalline silicon solar cell performances.[76] In the PERC cells, a trade-off is obtained by reducing the contact surface using small local openings in the backside dielectrics. The problem is therefore reduced but not eliminated. To further, improve silicon solar cell performances it is necessary to eliminate any chemical structure introducing recombination centers in the device.

A very efficient solar cell technology, providing both optimal light absorption and long carrier lifetime, is the silicon Heterojunction Technology (HJT). HJT solar cells consist of a core of crystalline n-doped silicon sandwiched between two layers of amorphous hydrogenated silicon (Figure 11). These layers are composed by a sequence of doped (n on one side and p on the opposite) and intrinsic silicon amorphous silicon to provide

both the passivation and the required contact selectivity. Finally, a transparent conductive oxide is introduced to activate lateral conduction and allow charge extraction by the metal grid.
Therefore, silicon heterojunction technology (HJT) combines the advantages of monocrystalline solar cells with the excellent passivation of hydrogenated amorphous silicon (a-Si:H). The cell takes benefit from the high band gap (Eg) of the hydrogenated amorphous silicon (Eg =1.75 eV), while ensuring a good passivation of the surface states at the heterojunction with the crystalline silicon (Eg =1.15eV). With respect to most common approaches (Al-BSF and PERC), the passivation of HJT solar cell is a very efficient solution. This is ensured by the non-metallic contacts (a-Si:H) acting simultaneously as selective contact and passivating layer.[77]

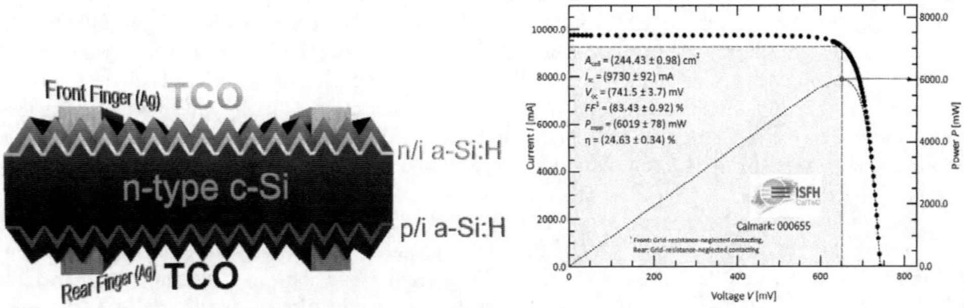

Figure 11. (left) Schematic of HJT solar cell. N-type crystalline wafer is sandwiched between amorphous intrinsic and doped silicon to improve passivation. ITO layers and metal grids allow to extract charges from the device. (Right) Record performances of a Si HJT solar cell, achieving an efficiency of 24.63%. The cell has been realized within a research collaboration between Enel Green Power and the INES - LITEN Institute (Institut Nationalle de L'Energie Solaire - Commissariat à l' Ènergie Atomique aux Ènergies Alternatives). The result achieved in January 2020 has been certified by the Cal-Tech Lab of the German ISFH (Institut für Solarenergieforschung in Hamelin). The cell is a prototype that is fabricated with standard materials and industrial processes and can be transferred in manufacturing in the next future. The research activities involve the most important European research and technology centers, within the framework of the European Horizon 2020 Project named "Ampere".

A fundamental characteristic for HJT cells is that the metallic contacts, which are usually source of recombination centers when contacting silicon, are electronically separated from the silicon active part that absorbs the light. This occurs due to both the high band-gap amorphous silicon layer and transparent conducting oxide (TCO). The TCO is a degenerate (Fermi level in the conduction band) and transparent n-type semiconductor used as non-metallic contact. Usually, the TCO is composed by an alloy of indium oxide and tin oxide, known as ITO. The separation of the crystalline core from the metallic contacts leads to record Voc values, owing to the absence of contact opening and direct junction between the metal and the silicon. Because of this characteristic, the HJT cell achieves higher efficiencies than conventional crystalline silicon technologies, with typical efficiency between 23% and 25%, and have more margins to approach the theoretical Shockley- Queisser limits of silicon solar cells. Up to date, HJT solar cells own the energy conversion efficiency between silicon solar cells technologies.[19]
In HJT cells, the photo-generated charges can flow slowly enough through the passivation layer to build-up a high voltage and at the same time they are fast enough to avoid recombination. The passivation layer acts as a semipermeable membrane for charge carriers extraction.[78] HJT concept has a certain affinity with the metal-insulator-semiconductor structure (MIS), which is based on quantum tunneling effect through a thin dielectric layer. To take advantage from the high passivation level of HJT structure, it is fundamental to use monocrystalline silicon with high minority-carrier diffusion length, in the millisecond range. This can be achieved with n-type silicon.
From the industrial viewpoint, with HJT it is possible to improve the passivation of silicon solar cells with very effective processes. These processes are conceptually simple and require quite low temperatures.

IOP Conf. Series: Materials Science and Engineering 1265 (2022) 012002 doi:10.1088/1757-899X/1265/1/012002

Moreover, thanks to the high device symmetry, fewer process steps are required if compared to PERC technology. Figure 12 shows the comparison of manufacturing process flows of PERC and HJT, respectively. Because of these important advantages, PV module manufacturer interest in the technology is increased in the last years.
The HJT solar cell manufacturing can be schematized as following. After the visual inspection of silicon wafers, texturing and cleaning processes are carried-out in wet bench using sequential etching in KOH, rinse in water and etching in HCl and HF. Afterwards, the hydrogenated amorphous silicon layers are deposited with plasma enhanced chemical vapor deposition (PECVD). The deposition is performed with the sequence i/n on the front face and i/p on the rear side (i: intrinsic, n: n-doped, p: p-doped).
The technique allows a conformal covering of the crystalline silicon textured surface with amorphous silicon. The thickness of the amorphous layer can be accurately controlled during the deposition. The thickness of the intrinsic amorphous silicon is of few nanometers while the doped layers are of ⍰ 10nm. The deposition process is performed at temperatures below 200°C. The a-Si layer also contains a high percentage of hydrogen (⍰ 10%) contributing to the definition of the optical and electrical properties, including the band-gap value. Besides, hydrogen saturates the silicon dangling bonds at the interface,[77] thus, drastically reducing the density of trap states.
After the deposition of the a-Si:H layers, the TCO layers are deposited via sputtering, a vacuum deposition technique (Physical Vapor Deposition, PVD). From the physical point of view, sputtering relies on the generation of a plasma to erode a target substrate. The sputtered and ionized molecules are than directed towards the substrate through an electromagnetic field to form a thin film of the desired material. In HJT manufacturing, the TCO deposition occurs with ITO target bombarded with Argon and or oxygen ions. For HJT, the ITO targets are generally constituted for more than 95% by Indium Oxide and for less than 5% by Tin Oxide.
The formed ITO layer acts in the final device as a conducting transparent layer with a sheet resistance ranging between 5 and 100 □/square. Due to its transparency, the film operates as a window allowing the light to arrive efficiently in the silicon absorber layer. Moreover, ITO films act as Ohmic contacts, allowing horizontal charge transport for the extraction of the charges from the solar cell.[79] To collect and extract the photogenerated charges effectively, a metallic grid is needed. The grid, made of silver, is printed on the TCO using the screen printer technique. Though the technology processes are quite sophisticated, HJT technology is characterized by a simple process flow with a small number of steps, significantly lower than those of conventional technologies like PERC. HJT solar cell has a very symmetric structure with efficient passivation both on the front and the backside of the cell thanks to the amorphous silicon layers that can be industrially upscaled.[80–82] Thanks to its high symmetry, it also exhibits the highest bifacial ratio. Typical HJT cell own bifacial ratio exceeding 90% and can approach the 95%.[71] Besides superior performances in terms of efficiency and bifacial factor, HJT cells show other advantages that are mainly observed in the average electricity produced by a PV plant. The better surface passivation enables a higher control of solar cell performances at high temperature operations, due to a reduced thermal degradation factor, compared with conventional crystalline cells.[71] Moreover, there is a better behavior at low irradiance and grazing angles due to the optical characteristics of the TCO. The behavior of HJT solar modules is even better with time because of a sort of regeneration effect leading to increase the performances after exposition to the light.[83,84] HJT resilience is related to the winning combination of high-quality n-type silicon and passivation effects of amorphous silicon on both surfaces. In addition, the use of n-type wafer for HJT cells avoid the drawback of LID and LeTID since the absence of boron helps eliminate Voc degradation during exposure to the light.

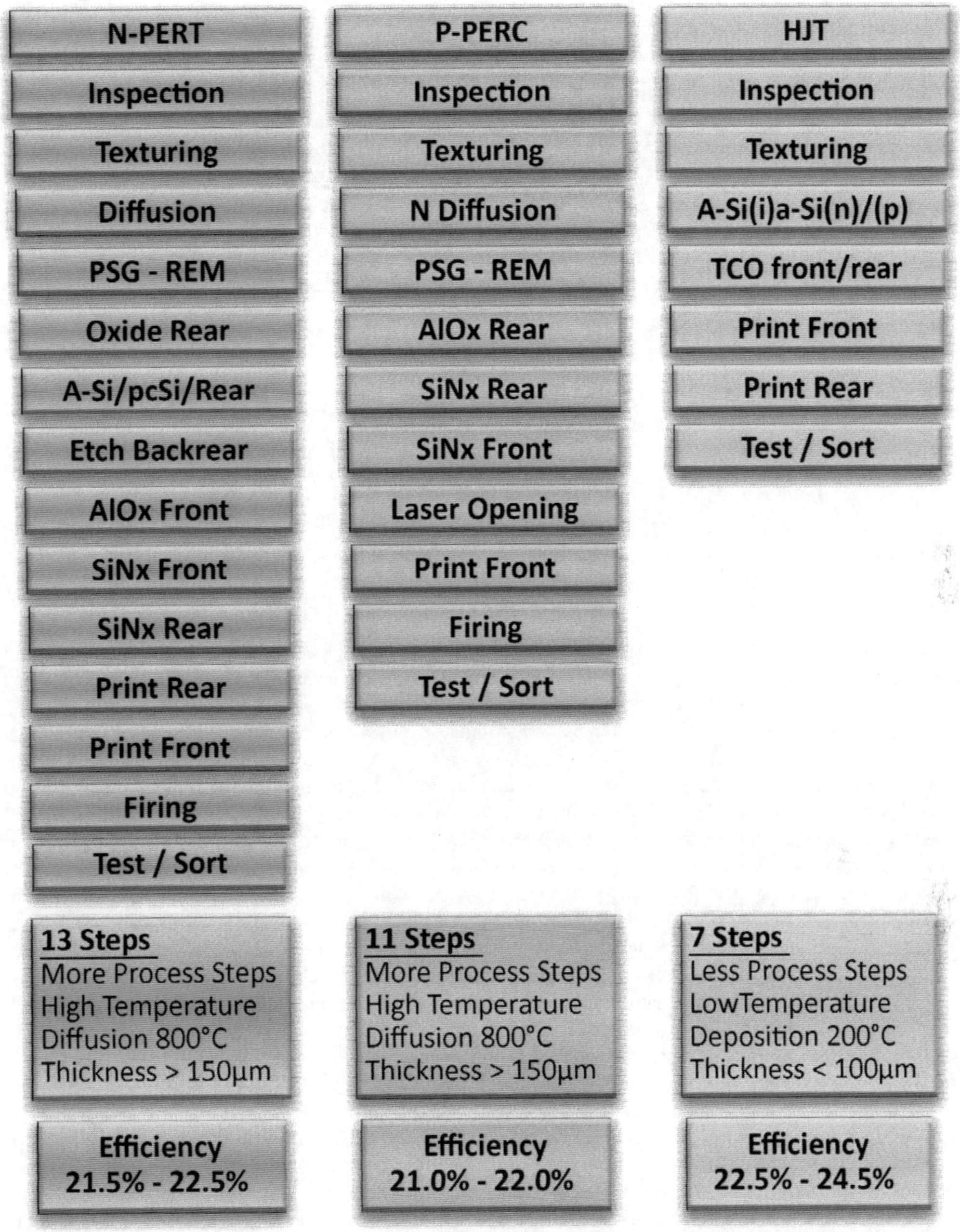

Figure 12: Comparison between process flows for N-PERT, P-PERC and HJT solar cell fabrication, respectively. HJT has a lower number of process steps. This facilitates the use of automated operations allowing to reduce the overall manufacturing costs.

In PECVD, a plasma generated with radio-frequency electromagnetic fields allows to dissociate the molecules in the precursor gases to obtain the deposited film. The deposition occurs under vacuum enabling also highly pure and clean film formation. In HJT manufacturing, silane (SiH4) mixed with hydrogen (H2) is employed as precursor to form the hydrogenated amorphous silicon layers. Phosphine (PH3) and di-borane (B2H6) are employed as n-doping p-doping precursors respectively for the amorphous silicon layers.

HJT Modules Characteristics	Advantages with respect to Mainstream
Less process steps	Lower manufacturing transformation costs
Low temperature process (<200°C)	Better cell electronic properties. Possibility of reducing silicon thickness and hence reducing silicon cost.
Excellent surface passivation and selective contacts	Higher efficiency due to higher Voc
Low thermal degradation coefficient thanks to the excellent surface passivation	> 8% energy generation in warm climates
Well insulated solar cell thanks to the presence of a TCO continuous layer on both faces (no openings and no direct contact of metal with silicon)	Better reliability with a degradation even lower than 0.25%/year
High architecture symmetry, bifacial factor of 90-92%	Higher bifacial gain (even more than 95% in best cases). Higher energy gain up to +25% with favorable albedo conditions.

Table 2: shows the advantages of HJT compared with mainstream PERC technology. HJT technology has the best numbers for reducing energy costs and can became the technology of reference in the next future.

4.6 Selective Contacts

In the case of crystalline silicon technologies, achieving a good contact structure has become a fundamental objective to improve solar cell performances and approach the Shockley-Queisser theoretical limits. To develop such efficient contact structures, it is necessary to find processes and materials able to reduce recombination losses associated to the contacts. In particular, the ideal contact should simultaneously passivate the semiconductor surface and selectively extract charge carriers. As previously discussed, the direct junction of metal contacts and silicon leads to charge recombination and therefore should be avoided. At the same time, as shown for HJT solar cells, the efficient selective extraction of the charges at the two opposite sides of the device further decreases the recombination of the charges and increases the Voc of the device. As a result, the ideal contact should consist of materials interposed between silicon and the metallic electrodes, able to successfully suppress the recombination at the interface states while acting simultaneously as semipermeable membranes which select the suitable charge carriers (Figure 13). In solar cells, holes are extracted from the p-type side of the p-n junction whereas the electrons are extracted from the n-type side. However, in real devices also minority charge carriers are present close the electrodes (holes in the n-type layer and electron in the p-type side). As a result, charges can recombine in the traps existing at the interfaces of the contact region of the device. Hence, to achieve high performances it is necessary that only the majority carriers are in correspondence of the contact. HJT solar cells use amorphous silicon layer to exploit charge carrier selection in an effective way, thus outperforming conventional solar cells architectures. It is possible to even improve HJT performances by using materials with high work-function (WF) and higher band-gap to create, in correspondence of the electrodes, a higher offset in the conduction band (for electrons) or in the valence band (for holes).[78] Such a band structure can keep away the minority charge carriers from the electrodes allowing only the majority charge carriers to be extracted. Furthermore, since the selective contact and the absorber own different WF, a band bending can be observed in the absorber band structure. This leads to the formation of an induced junction. A correct cell structure design can have a rectifying behavior similarly to a junction formed by n- or p-type doping of the absorber surface. Furthermore, selective contact design can yield higher overall voltage level. In fact, the use of selective contacts on top of the c-Si absorber not only provides a selective excess carrier extraction, but also enables good chemical passivation of the c-Si surface and low charge recombination in the selective layer.[85] Besides, the detrimental influence of very high carrier densities in the contact region, i.e. heavy doping effects like Auger recombination and band gap narrowing, is expected to be of minor importance. However, individuating new materials with the ideal properties is always a long a complex task. For this reason, the PV community has taken advantage of the high WF materials already adopted in organic electronics[86,87] investigating their use for silicon thin film and,[88,89] more recently, for c-Si absorbers.[90,91] Within the number of materials that could be used for this application (Figure 14) thermally evaporated molybdenum oxide (MoOx) revealed to be an interesting candidate to replace the p-type a-Si:H.[92] Although the material was already proposed back in 1980,[93] its application in highly efficient industrially compatible devices is still under investigation. In fact, while MoOx can outperform a-Si:H(p) in the as-deposited state, a crucial challenge is the formation of a hole barrier during low temperature annealing adversely affecting the FF and even the Voc.[94,95] Currently, this is a strong limitation towards the industrialization of such materials. It was recently shown that such behavior is related to the presence of hydrogen in the adjacent layers.[96] As an alternative to thermally evaporated films, MoOx formed by thermal atomic layer deposition (ALD) or plasma enhanced atomic layer deposition ALD have been investigated. [97–100] Though ALD is an interesting approach for the industrial deposition of ultra-thin films[101], ALD deposited MoOx solar cells currently still exhibit a low efficiency compared to their evaporated counterparts.[102]

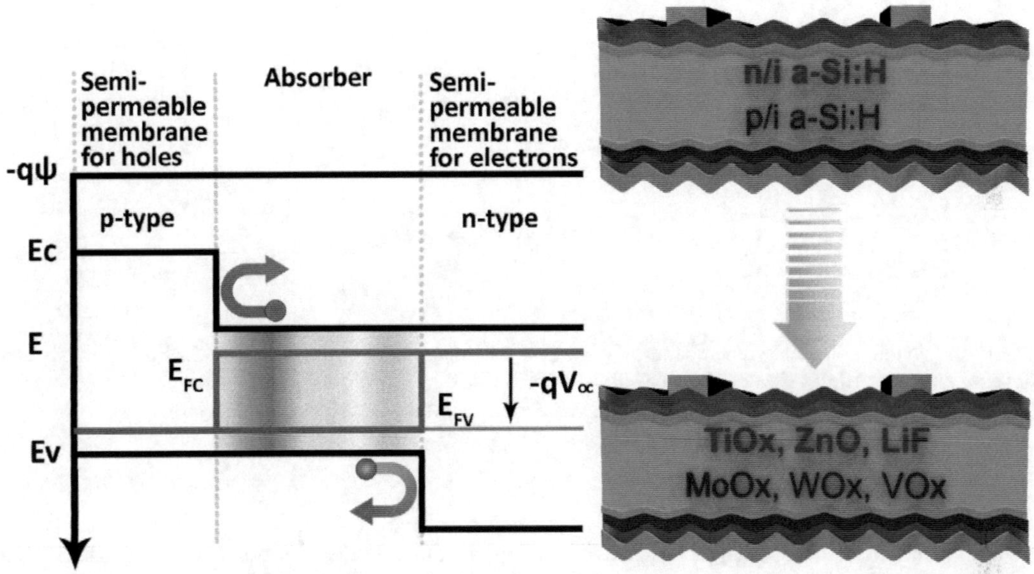

Figure 13: (Left) Ideal solar cell band structure with selective contacts for electrons and holes. In correspondence of the electrode in the p region, the electrons (in red) in the conduction band see a barrier that rejects them towards the opposite direction. Vice versa, the holes (in bleu) in the valence band encounter a barrier in the n region and are repelled in the opposite direction. Therefore, electrons can reach the n-type electrode without recombining in the p-type region. Analogously, holes can reach the p-type region and can be collected without recombining in the n-type electrode zone. (Right) structures with selective contacts based on amorphous silicon/ crystalline silicon (above), and improved selection structure for holes with MoO_x, WO_x o VO_x and for electrons with TiO_2 MgF_2, ZnO

Whereas MoOx represent a valid candidate to replace amorphous silicon as hole selection layer as p-type contact, TiOx could represents an interesting material for the replacement of the electron selective contact.[103,104] However, it is not easy to find deposition methods compatible with industrial processes at low cost, which can deposit thin layers at the interface without damaging the structure underlaying. Remote Plasma Physical Vapor Deposition (RPPVD) and Pulsed Laser Deposition (PLD) represent interesting industrial route for the deposition of such materials since both the approaches causes low surface damages that could be reproduced at industrial scale.

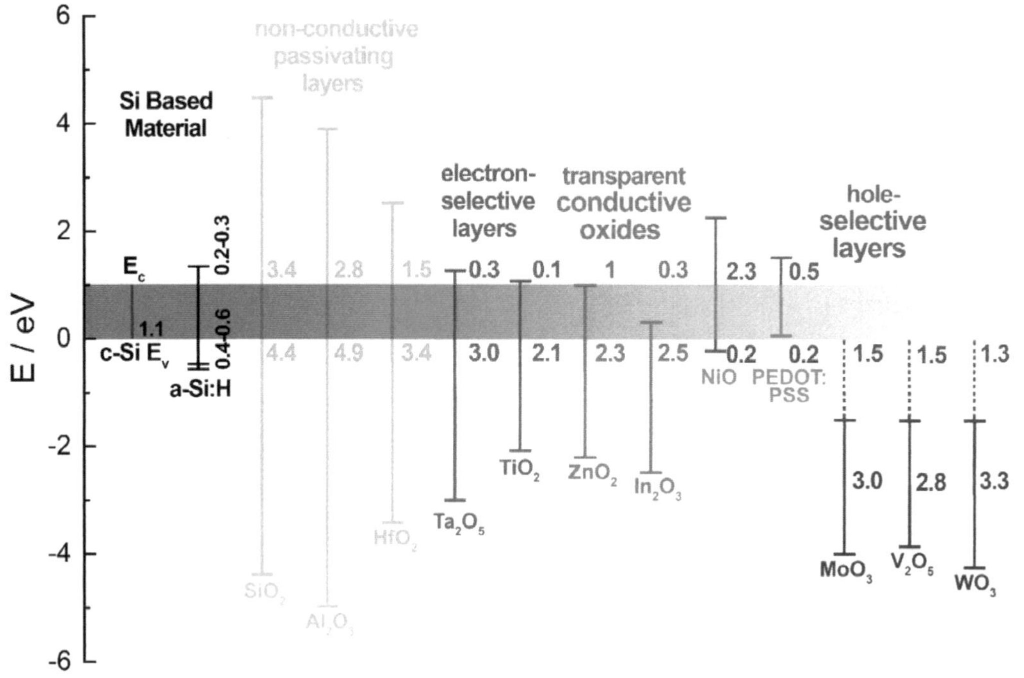

Figure 14: Conduction and valence band offsets of various materials with respect to c-Si. For materials that have an electron affinity greater than that of c-Si (as various TCOs, MoO3, V2O5 and WO3) the energetic difference between the conduction band of those materials and the c-Si valence band is shown.

4.7 Thin Crystalline Silicon

In a high efficiency crystalline silicon PV module, the wafer covers about 25-30% of the full module cost.[107] As a consequence, reducing the wafer thickness is an important tool for the reduction of the used material and hence of costs. Thinning the silicon wafer well below the industry-standard 180-160 μm, in principle reduces both manufacturing cost and capex, and accelerates economically sustainable expansion of PV manufacturing.[108] Capex and other costs related to wafer manufacturing occurring prior to the wafer formation through wire sawing scale down proportionally with the amount of silicon used per Watt of produced electricity. Today, technologies able to produce thin silicon wafers down to 100 μm thickness are already viable.[109]

Thickness reduction to 50 μm or less require production processes known as "kerfless".[110] Examples of kerfless processes are epitaxial mono-wafer[111] and directly-grown multi-wafer technologies[112]. kerfless wafers have not yet been adopted at a large scale, mostly because of a lack of market for thin wafers. With the inevitable trend of utilizing thinner and thinner wafers, one may ultimately turn to these viable kerfless technologies to extract the maximum silicon savings. From the PV cell and module fabrication processes viewpoint, thickness reduction implies less mechanically robust silicon wafers leading to wafer breakage during high temperature treatments. This aspect applies less to HJT cells thanks to the low thermal budget required (<200°C) enabling the use of thin wafers, with thickness of 100 μm or even below. Nevertheless, wafer thinning decreases the volume of absorber material and hence the ability to absorb light. Below 100 μm the short circuit current given from the integral of the absorbed photon flux starts to decrease. To compensate the current loss, it is necessary to increase the light capturing through light photon management approaches. The scope of these approaches is to increase the optical path of light or to increase the light trapping in the absorber material. There are different approaches relying on nanostructures realized on the silicon wafer.[105,113,114] Another technique could be the use of metallic nanoparticles enabling photons / plasmons coupling to maximize the absorbed light.[115–117] However, these technologies are not mature to be used as industrial standards.

Figure 15: Lower temperature processes as in the case of HJT solar cells enable the use of thinner crystalline silicon, even thinner than 100 µm (left). (right) The use of lower temperature and substrate thickness leads to more resilient devices.

4.8 Tandem Solar Cell

As discussed in the previous paragraphs, the Shockley-Queisser limit indicate that the maximum conversion efficiency of silicon solar cell in ideal conditions in about 29%. To calculate such upper conversion efficiency limit, several factors are considered such as the Auger recombination (in Auger recombination the energy of an electron, which transits from the conduction band to valence band is conserved by transferring the energy to another electron or hole), intrinsic losses, transparency of silicon to photons with energy lower Eg and thermalization losses for high energy photons generating high energy electrons and holes (hot carriers).
To overcome the limitations imposed by the S-Q limit, the use of multiple absorbers with different bandgaps allows to increase the theoretical efficiency limit. In principle the use of two absorbers could already allow to achieve efficiencies above 40%.[105] Moreover, it is theoretically possible to use an infinite number of superimposed solar cells to achieve efficiency above the 80%.[105] Practically, the most effective approach consists in employing two sub-cells to obtain the best trade-off between light absorption and losses reduction. We refer to this kind of solar cells architectures employing multiple absorbers as "Tandem solar cells". Tandem solar cells consist in the connection of two or more superimposed solar cells with different band gap (Figure 16). Commonly, the sub-cell facing the sun is referred as top cell which is stacked on top of a bottom cell. To allow the light to pass through all the stacked cells, the bandgaps are engineered decreasing from the top cell to the bottom one. In this architecture, the top cell absorbs the blue fraction of light (high energy photons) and is transparent to lower energy photons. The low energy photons are then absorbed by the underlying cells.

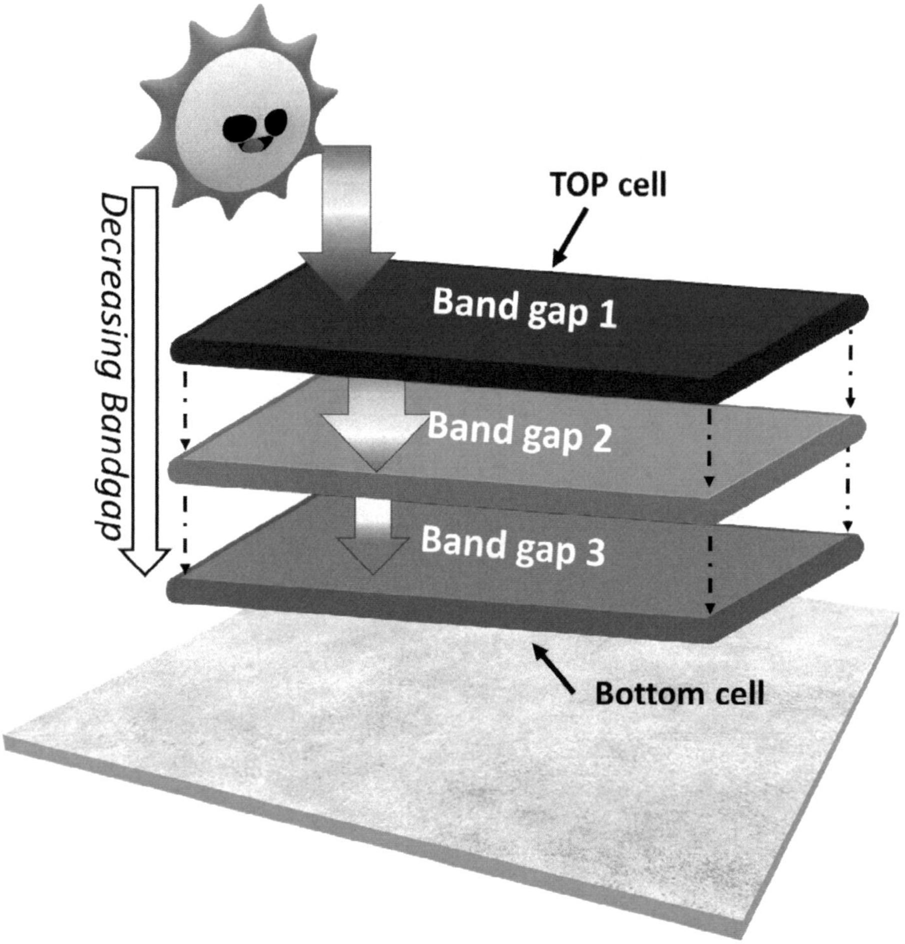

Figure 16: Scheme of tandem solar cells where solar cells with three different bandgaps are stacked one above the other.

These architectures have already shown an optimal use of the solar spectrum allowing to obtain very high power conversion efficiencies.[106] In tandem stacks, the sub-cells can be connected in series or parallel. In series connections, the current of the system is determined by the lowest current that is generated by the less performing cell, while the output voltage is equal to the sum of the two sub-cells Voc:

$$V = Voc_{Top} + Voc_{Bottom}$$

On the contrary, in case of parallel connections it is necessary to achieve voltage matching while the current obtained is the sum of the devices Jsc.

$$J = Jsc_{Top} + Jsc_{Bottom}$$

Another key aspect of tandem solar cells is their architecture. Tandem cells can be assembled in two different configurations: 2-Terminals (referred as 2T) and 4-Terminals (referred as 4T) (Figure 17). In case of 2T cells (also referred as monolithic approach) the tandem cell is conceived as unique device depositing the top cell directly on top of the bottom cells. In this approach, the two sub-cells are connected in series providing a device with only two electrodes. On the contrary, the 4T approach consists in stacking two different devices each one owning a positive and a negative electrode. In this case, both series and parallel connections.

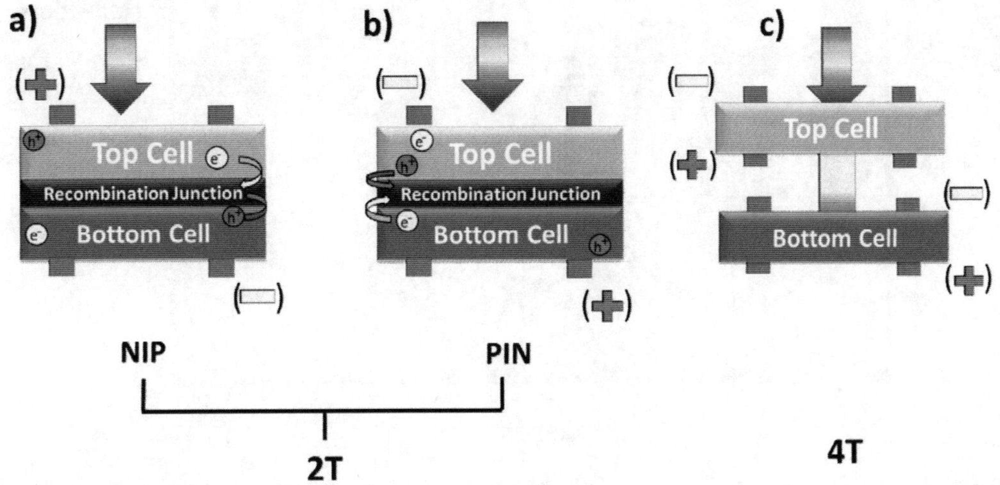

Figure 17: Scheme of 2T and 4T tandem solar cells stack.

In general, cells with low bandgaps generate higher current as they absorb a higher fraction of the solar spectrum. On the contrary, cells with wide bandgaps provide higher voltage outputs and lower currents. The optimum condition is achieved at intermediate values, in the 1.0 -1.6 eV range. Silicon solar cells (Eg =1.1eV) and CIGS cells (1.08eV) are close to the lower part of the range, GaAs cells (Eg=1.45eV) and CdTe (Eg =1.5eV) own Eg in the higher section of the range.

In 4T tandem structures, both the cells can operate independently at their maximum power points, with distinct tracking systems. This reduces the limitations on the choice of the top cell band gap and makes the system less sensitive to the spectral variations.[107] Consequently, the 4T structure can reach higher efficiencies, with top cells with bandgap that can vary between 1.6eV and 2.0eV, with an optimal value of 1.8eV, when a silicon cell is used as bottom cell (see Fig.18).

However, the 4T configuration brings several problems related to the transparency of the contacts blocking the light path towards the bottom cell. Furthermore, the 4T configuration implies higher costs related to the necessity to duplicate the amount of glass, encapsulant and electronics required.

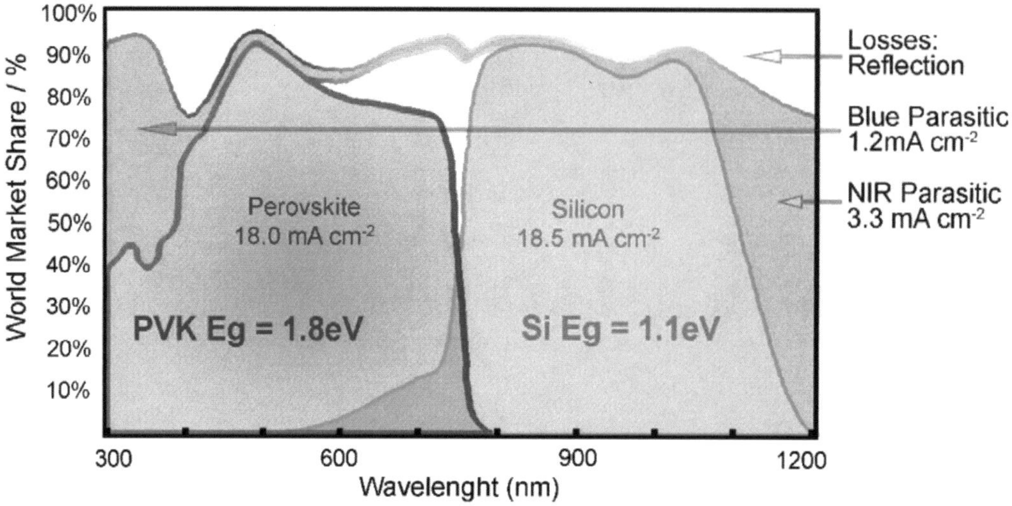

Figure 18: External Quantum efficiency (EQE), red and blue curves, and optical losses (1-R, where R is the cell reflectance), in gray, with a tandem with a perovskite (with 1.8 eV band gap) top cell and a silicon bottom cell. The top cell uses the blue component of the solar spectrum, whereas the bottom cell uses the red component [34].

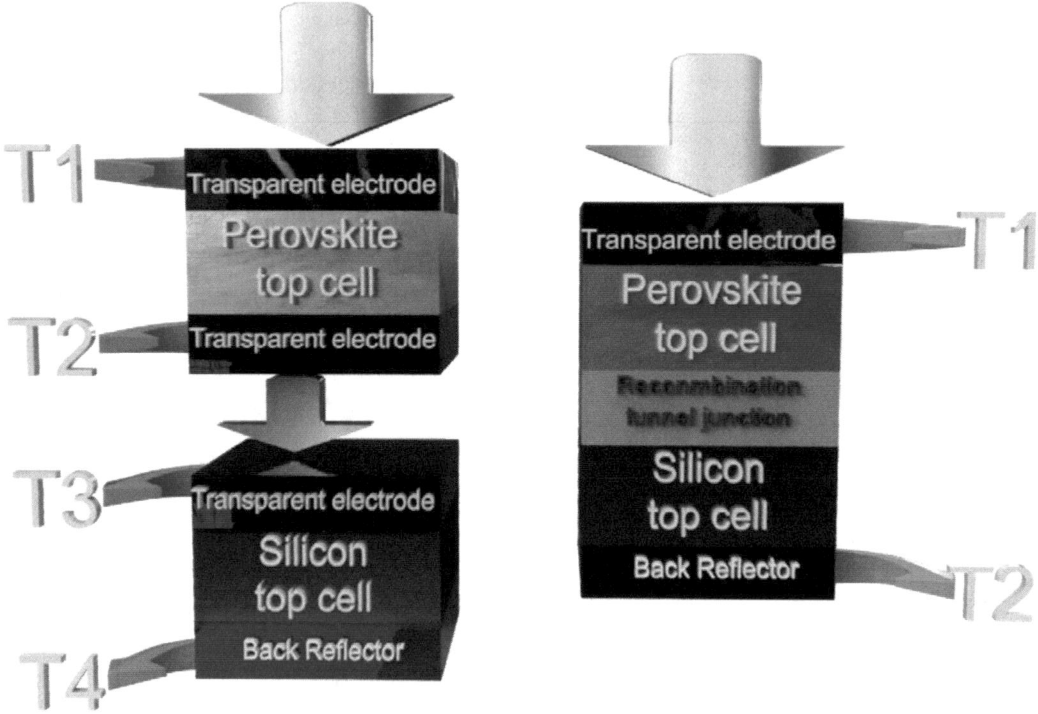

Figure 19: (Right) Four terminals tandem cell. (Left) Two terminals monolithic tandem cell [34].

On the other hand, in the monolithic two terminals (2T) approach the top cell is deposited directly on the bottom cell. With respect to the 4T structure, a 2T Tandem requires only one transparent electrode thus reducing the impact of parasitic optical absorption, the number of process steps and the fabrication costs. The interface layer between the two cells is named "recombination junction" and allows the current flowing between the two cells with recombination mechanism or with quantum tunneling.[108] This layout imposes the series connection of the top and bottom cells. With this configuration, less materials such as glass, electrodes, encapsulant and electronics are required.[109] For this reason, although this technology represents a more disruptive approach and would arrive to the market less smoothly, the industrial interest on 2T tandems is very high.
However, also 2T tandems can have some drawbacks, mainly consisting in technical and technological obstacles. Firstly, the two cells are connected in series requiring the matching of the current deriving from the two sub-devices. This limits the range of optimal bandgap for the functioning of the device.[107,110] Secondly, the top cell must be built directly on top of the bottom cell. Therefore, using crystalline silicon bottom cells, the texturization of the substrate inevitably forces towards the use of conformal deposition techniques to avoid shunts.

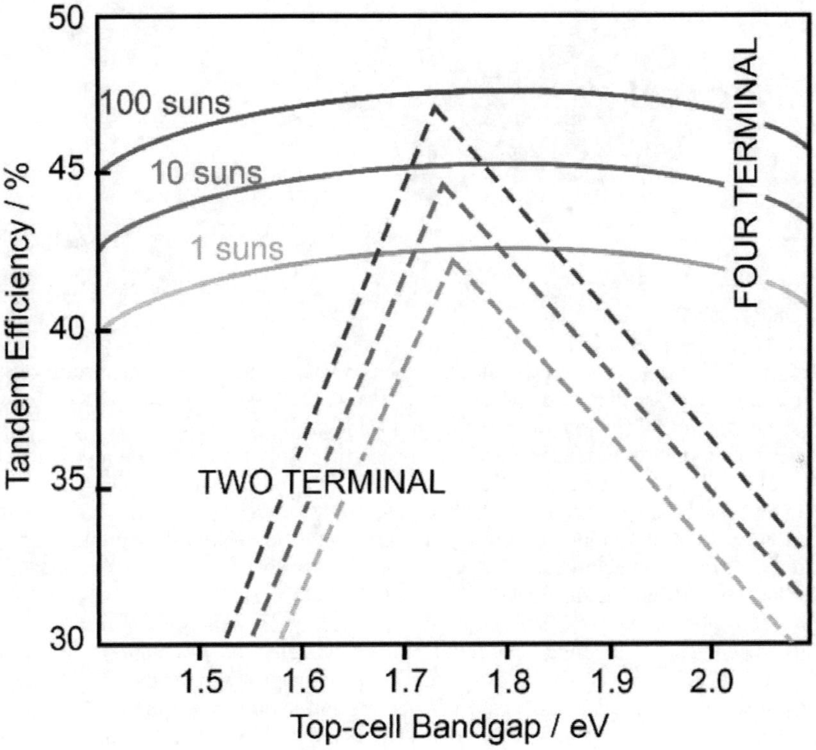

Figure 20: Efficiency theoretical limit, calculated as a function of the top cell band-gap, for a tandem cell with four terminals and with two terminals, respectively [34].

Using the detailed balance principle, a theoretical limit of about 40% can be attained for a perovskite/silicon tandem cell. In the case of the 2T tandem a sharp window is found and the efficiency decreases abruptly as one moves from the ideal band gap of the top cell, because of the constrain of matching the currents of top and bottom cell. [107]
For a Tandem approach with two junctions, the silicon solar cells are quasi-ideal bottom cells, because of their low Eg (1.1 eV), relatively high Voc (that for HJT can even exceed 750 mV, corresponding to the highest values for crystalline silicon based solar cells) and high efficiency. On the other hand, it is quite difficult to find an ideal counterpart as top cell.

Figure 21: ABX3 structure characterizing the perovskite, with the octahedral BX6 and the larger cation A positioned in a cubic-octahedral site.

Solar cells based on III-V compounds semiconductors, such as GaAS and InP, are interesting for they high efficiency[19] and almost ideal band gap value for a top cell. However, high fabrication costs of such materials do not facilitate the widespread diffusion for terrestrial applications.

An interesting approach to industrial tandem application can come from the exploitation of perovskite solar cells. Perovskite solar cells already demonstrated to be a promising top cell in combination with crystalline silicon based bottom solar cell.[109] The name Perovskite is related to the name of the Russian mineralogist Lev Perovskite, who was the first to identify the crystalline structure ABX3. In this formula A represents a monovalent cation (occupying a cubic-octahedral site shared with twelve X anions), B is a smaller bivalent cation (stabilized in an octahedral site shared with six anions X) and X is an oxygen or a halogen atom (Figure 21). Historically, perovskite structures have been studied mainly as oxides with ferroelectric[111] or superconducting properties[112]. Halogen perovskites have attracted poor attention until it has been discovered their potential as absorbers in solar cells. The first experiments reported the use of CH3NH3PbX3 as absorber in dye sensitized solar cells.[113,114] However, the cells exhibited extremely short life-time until Park and co-workers were able to achieve the device improved stability while employing a solid hole conductor.[115] Since then, perovskite solar cells have seen an incredibly fast increase in their stability and efficiency up to the most recent record above 25%.[18] One of the more interesting aspects of using ABX3 perovskite absorbers, is their versatility in terms of bandgap, which can be tuned varying the halogen composition (X= I, Br, Cl),[116–118] the A cation (generally Methylamonium-MA, Formamidinium-FA or Cesium)[119–121] or the B heavy metals (Sn or Pb).[122] The most investigated perovskite is probably the MAPI (CH3NH3PbI3) due to its easy preparation and ideal bandgap. Nevertheless, several studies showed that FA based perovskite are more stable and would be better candidates for industrial devices.[123,124] Further improvements can be achieved carefully engineering the perovskite composition using both FA, MA and Cs simultaneously as A cation.[125] Another critical aspect concerns the photoinduced phase segregation in case of multiple halide reducing the overall efficiency and stability of the perovskite with time. It was shown that halide segregation impact more MA perovskites compared to FACs based perovskites.[126]

Perovskite cells can be fabricated in several ways. The most common technique is spin coating antisolvent method. In this approach a solution containing the precursor of the perovskite is deposited in the center of a rotating substrate. The centrifugal force will than spread the material all over the surface. In a second step an antisolvent is deposited on top of the substrate. The antisolvent strips-away the remaining solvent causing the formation of the perovskite. The phase of interest is than formed with a thermal treatment. However, although this method is widespread in laboratories all around the world, it is poorly compatible with industrial standards.[127] Other solution techniques, such as slot-die coating or chemical vapor deposition and co-evaporation, have the potential to be industrially scaled-up. The performances of the perovskite solar cells obtained with such techniques still lay behind the record obtained by spin coating. Recent studies suggest that the co-evaporation allows to achieve higher thermal stability[128] while retaining better performances increasing the device size.[129] For this reason evaporation may represent a key route towards the commercialization of perovskite technologies. It is also noteworthy that, so far, most of the good performances and stability are obtained only in the laboratory under controlled conditions (crystallization processes achieved in glovebox inert environment with N2 atmosphere).

The perovskite technology still retains two major weak points to be efficiently scaled up and commercialized both as a single junction and as a tandem device: i) difficulty in obtaining high uniformity in industrially relevant substrate sizes ii) poor stability compared to state-of-the-art silicon-based technologies. To transfer the perovskite technology to consolidated industrial manufacturing processes it is mandatory the development of more robust and reproducible fabrication processes. Therefore, it is fundamental to protect efficiently the perovskite cell against environmental agents (including moisture and oxygen), light and voltage induced degradation. Despite important studies have shown that perovskite solar cells could drastically increase their stability and robustness through encapsulant engineering,[130] perovskites are still in an experimental stage and exhibit several drawbacks for scaling up the processes (such as high performances achieved only on cells of few cm2).[129] Another limiting aspect for the widespread diffusion of perovskites on commercial scale involves the fact that lead (Pb) is a key component of the perovskite structure. At the same time Pb introduces serious problems of toxicity in the solar cell manufacturing and handling. To allow perovskite solar cells to be used in an industrial tandem structure it is necessary to solve the disadvantages related to scalability, non-toxicity and reliability. On the other hand, the scientific community is enthusiastically concentrated to find solutions to the mentioned drawbacks and every few months significant results are achieved.

Although those limits, Perovskite solar cells represent an ideal candidate as top cell in tandem structures. Several studies showed in few years the route towards efficiencies higher than the record obtained in silicon single junctions, closely approaching the 30% efficiency.[109,131] At the same time the scientific community is showing that several routes are possible to overcome the perovskite limits. Several groups showed that modifying the pyramid texturing process it is possible to deposit the top cell directly on the silicon solar cells with several techniques such as blade-coating,[132] hybrid vacuum/ liquid deposition,[133] or solution processing.[134] Albrecht and co-workers showed that the stability of the tandem structure can be drastically increased with an accurate selection of the materials composing the stack.[109] At the same time, the engineering of the perovskite layer can lead to significant increase in the device life-time and stability.[131,135] Di Carlo and co-workers showed that it is possible to assemble 2-T devices mechanically stacking two sub-cells separately prepared. This approach allows to circumvent the thermal budget constrains generally intrinsic in 2T devices while boosting of the device efficiency thanks to 2D materials integration in the various layers.[136]

5. Conclusions

The huge manufacturing volume increase of PV modules and the corresponding reduction of the costs have remarkably improved the perspectives for a fast transition towards a more sustainable future. Current research and the concentrated industrial effort suggest that silicon solar cell technology will continue its evolution reaching efficiency close to the theoretical limit of 29%. The utilization of advanced passivation schemes, such as the use of thin silicon films in HJT solar cells as well as the implementation of more efficient selective contacts, will facilitate such a transition. The increased efficiency will lead to a general cost reduction of electricity. The current panorama strongly indicates that the cost reduction of the PV generated electricity will be supported by the development of new PV module architectures able to minimize intrinsic electric and optical losses. To overcome the limit of silicon based solar cell technologies, the market will have to rely on new technologies. Current state-of-the art suggests that tandem architectures represent one of the more suitable candidates to achieve such results. In particular, the coupling of thin silicon bottom cell owing a consolidated industrial process with a top cell made with a low-cost thin film material appears to be one of the most industrially appealing technology. However, several challenges are still to be overcome to obtain reliable tandem technology to exploit industrially, such as finding suitable materials in terms of stability, non-toxicity and availability for mass scale. Continuing with the trend of market growth observed in the last years, photovoltaics will be more and more diffused and will succeed to become the electricity source at lower cost, providing a very low cost and sustainable solution for the challenge of decreasing the global warming. The initial steps of solar cell research have been driven by the need of providing energy to the satellites and by the launch of the new era of space exploration. In the next years, a big research effort, focused on new materials, PV modules design, reliability and product integration and manufacturing methodology will be required to address the challenges related to multi-terawatt scale PV deployment.

6. References

[1] Green M A 2005 Silicon photovoltaic modules: A brief history of the first 50 years *Prog. Photovoltaics Res. Appl.* **13** 447–55
[2] Green M A 2009 The path to 25% silicon solar cell efficiency: History of silicon cell evolution *Prog. Photovoltaics Res. Appl.* **17** 183–9
[3] Galluzzo F R, Zani P E, Foti M, Canino A, Gerardi C and Lombardo S 2020 Numerical modeling of bifacial PV string performance: Perimeter effect and influence of uniaxial solar trackers *Energies* **13**
[4] Kopecek R and Libal J 2021 Bifacial photovoltaics 2021: Status, opportunities and challenges *Energies* **14**
[5] Leonardi M, Corso R, Milazzo R G, Connelli C, Foti M, Gerardi C, Bizzarri F, Privitera S M S and Lombardo S A 2022 The effects of module temperature on the energy yield of bifacial photovoltaics: Data and model *Energies* **15**
[6] Anon Fraunhofer ISE
[7] Anon Solar Power Europe
[8] Fraunhofer Institute for Solar Energy Systems ISE with support of PSE GmbH Photovoltaics Report.
[9] ITRPV 2021 2021 *2021 ITRPV.pdf*
[10] Singh D, Chaudhary R and Karthick A 2021 *Review on the progress of building-applied/integrated photovoltaic system* vol 28 (Environmental Science and Pollution Research)
[11] Green M A 2019 Photovoltaic technology and visions for the future *Prog. Energy* **1** 013001
[12] Swanson R M 2006 A vision for crystalline silicon photovoltaics *Prog. Photovoltaics Res. Appl.* **14** 443–53
[13] Haegel N M, Atwater H, Barnes T, Breyer C, Burrell A, Chiang Y-M, De Wolf S, Dimmler B, Feldman D, Glunz S, Goldschmidt J C, Hochschild D, Inzunza R, Kaizuka I, Kroposki B, Kurtz S, Leu S, Margolis R, Matsubara K, Metz A, Metzger W K, Morjaria M, Niki S, Nowak S, Peters I M, Philipps S, Reindl T, Richter A, Rose D, Sakurai K, Schlatmann R, Shikano M, Sinke W, Sinton R, Stanbery B J, Topic M, Tumas W, Ueda Y, van de Lagemaat J, Verlinden P, Vetter M, Warren E, Werner M, Yamaguchi M and Bett A W 2019 Terawatt-scale photovoltaics: Transform global energy *Science (80-.).* **364** 836–8
[14] Vartiainen E, Masson G and Breyer C PV LCOE in Europe 2015-2050 *31st European Photovoltaic Solar Energy Conference* (Hamburg, Germany)
[15] ITRPV 2021 2021 *International Technology Roadmap for Photovoltaic 2021*
[16] Shockley W and Queisser H J 1961 Detailed balance limit of efficiency of p-n junction solar cells *J. Appl. Phys.* **32** 510–9
[17] Henry C H 1980 Limiting efficiencies of ideal single and multiple energy gap terrestrial solar cells *J. Appl. Phys.* **51** 4494–500
[18] Min H, Lee D Y, Kim J, Kim G, Lee K S, Kim J, Paik M J, Kim Y K, Kim K S, Kim M G, Shin T J and Il Seok S 2021 Perovskite solar cells with atomically coherent interlayers on SnO2 electrodes *Nature* **598** 444–50
[19] Green M A, Dunlop E D, Hohl-Ebinger J, Yoshita M, Kopidakis N and Hao X 2022 Solar cell efficiency tables (version 59) *Prog. Photovoltaics Res. Appl.* **30** 3–12
[20] Zulehner W 1983 Czochralski growth of silicon *J. Cryst. Growth* **65** 189–213
[21] Shimura F 1989 Semiconductor silicon crystal technology 426
[22] Green M A, Dunlop E D, Hohl-Ebinger J, Yoshita M, Kopidakis N and Hao X 2021 Solar cell efficiency tables (Version 58) *Prog. Photovoltaics Res. Appl.* **29** 657–67
[23] Guerrero-Lemus R, Vega R, Kim T, Kimm A and Shephard L E 2016 Bifacial solar photovoltaics - A technology review *Renew. Sustain. Energy Rev.* **60** 1533–49
[24] Park N G and Zhu K 2020 Scalable fabrication and coating methods for perovskite solar cells and solar modules *Nat. Rev. Mater.* **5** 333–50
[25] Meillaud F, Boccard M, Bugnon G, Despeisse M, Hänni S, Haug F J, Persoz J, Schüttauf J W, Stuckelberger M and Ballif C 2015 Recent advances and remaining challenges in thin-film silicon photovoltaic technology *Mater. Today* **18** 378–84
[26] Bartlome R, Strahm B, Sinquin Y, Feltrin A and Ballif C 2010 Laser applications in thin-film photovoltaics *Applied Physics B: Lasers and Optics* vol 100 pp 427–36
[27] Berginski M, Hüpkes J, Schulte M, Schöpe G, Stiebig H, Rech B and Wuttig M 2007 The

effect of front ZnO:Al surface texture and optical transparency on efficient light trapping in silicon thin-film solar cells *J. Appl. Phys.* **101**
[28] Isabella O, Krč J and Zeman M 2010 Modulated surface textures for enhanced light trapping in thin-film silicon solar cells *Appl. Phys. Lett.* **97** 101106
[29] Zhang W, Paetzold U W, Meier M, Gordijn A, Hüpkes J and Merdzhanova T 2014 Thin-film Silicon Solar Cells on Dry Etched Textured Glass *Energy Procedia* **44** 151–9
[30] Li Z, Zhao Y, Wang X, Sun Y, Zhao Z, Li Y, Zhou H and Chen Q 2018 Cost Analysis of Perovskite Tandem Photovoltaics *Joule* **2** 1559–72
[31] Mandelkorn J and Lamneck J H 1973 A new electric field effect in silicon solar cells *J. Appl. Phys.* **44** 4785–7
[32] Fossum J G and Burgess E L 1978 High-efficiency p+-n-n+ back-surface-field silicon solar cells *Appl. Phys. Lett.* **33** 238–40
[33] Kim J, Lee S, Chowdhury S and Yi J 2022 A Brief Review of Passivation Materials and Process for High Efficiency PERC Solar Cell *Trans. Electr. Electron. Mater.* **23**
[34] Ye F, Li Y, Jia X, Guo H, Wang X, Ding J, Yuan N and Feng Z 2019 Optimization of phosphorus dopant profile of industrial p-type mono PERC solar cells *Sol. Energy Mater. Sol. Cells* **190** 30–6
[35] Fischer G, Strauch K, Weber T, Muller M, Wolny F, Schiepe R, Fulle A, Lottspeich F, Steckemetz S, Schneiderloechner E, Stegemann K H and Neuhaus H 2014 Simulation based development of industrial PERC cell production beyond 20.5% efficiency *Energy Procedia* vol 55 (Elsevier Ltd) pp 425–30
[36] Thibert S, Jourdan J, Bechevet B, Chaussy D, Reverdy-Bruas N and Beneventi D 2014 Influence of silver paste rheology and screen parameters on the front side metallization of silicon solar cell *Mater. Sci. Semicond. Process.* **27** 790–9
[37] Siraj S, Tahir S, Ali A, Amin N, Mahmood K and Manzoor A 2021 Assessing the Impact of Front Grid Metallization Pattern on the Performance of BSF Silicon Solar Cells *Silicon* **13** 4237–45
[38] Ju M, Lee Y J, Lee J, Kim B, Ryu K, Choi K, Song K, Lee K, Han C, Jo Y and Yi J 2012 Double screen printed metallization of crystalline silicon solar cells as low as 30 μm metal line width for mass production *Sol. Energy Mater. Sol. Cells* **100** 204–8
[39] Swatowska B, Stapinski T, Drabczyk K and Panek P 2011 The role of antireflective coatings in silicon solar cells *Opt. Appl.* **XLI** 487–92
[40] Stapinski T and Swatowska B 2006 Amorphous hydrogenated silicon–carbon as new antireflective coating for solar cells *J. Non. Cryst. Solids* **352** 1406–9
[41] Zighed L and Mahdjoub A 2010 Comparative study between various antireflective coatings for solar cells applications *IOP Conference Series: Materials Science and Engineering* vol 13 (Institute of Physics Publishing)
[42] Joshi A, Khan A and Sp A 2019 Comparison of half cut solar cells with standard solar cells *2019 Adv. Sci. Eng. Technol. Int. Conf. ASET 2019* 1–3
[43] Guo S, Singh J P, Peters I M, Aberle A G and Walsh T M 2013 A quantitative analysis of photovoltaic modules using halved cells *Int. J. Photoenergy* **2013**
[44] Xia L, Chen J, Liao K, Huang L, Li Q and Luo X 2020 Influence of laser cutting conditions on electrical characteristics of half-size bifacial silicon solar cells *Mater. Sci. Semicond. Process.* **105**
[45] Mittag M, Pfreundt A, Shahid J, Wöhrle N and Neuhau D H 2019 Techno-Economic Analysis of Half Cell Modules - The Impact of Half Cells on Module Power and Costs *36th EU PV Solar Energy Conference and Exhibition* (Marseille)
[46] Gérenton F, Eymard J, Harrison S, Clerc R and Muñoz D 2020 Analysis of edge losses on silicon heterojunction half solar cells *Sol. Energy Mater. Sol. Cells* **204**
[47] Baliozian P, Munzer A, Lohmuller E, Nair A, Fellmeth T, Wohrle N, Hoffler H, Spribille A and Preu R 2021 Thermal Laser Separation of PERC and SHJ Solar Cells *IEEE J. Photovoltaics* **11** 259–67
[48] Munzer A, Baliozian P, Steinmetz A, Geipel T, Pingel S, Richter A, Roder S, Lohmuller E, Spribille A and Preu R 2021 Post-Separation Processing for Silicon Heterojunction Half Solar Cells with Passivated Edges *IEEE J. Photovoltaics* **11** 1343–9
[49] Tonini D, Cellere G, Bertazzo M, Fecchio A, Cerasti L and Galiazzo M 2018 Shingling technology for cell interconnection: Technological aspects and process integration *Energy Procedia* vol 150 (Elsevier Ltd) pp 36–43
[50] Foti M, Galiazzo M, Cerasti L, Sovernigo E, Gerardi C, Guglielmino A, Litrico G, Sciuto

M, Spampinato A, Ragonesi A, Rametta F, Canino A, Carbonaro A, Coco F, Di Stefano A and Bizzarri F 2021 Silicon Heterojunction Solar Module using Shingle interconnection *Conf. Rec. IEEE Photovolt. Spec. Conf.* 1092–5

[51] Lindroos J and Savin H 2016 Review of light-induced degradation in crystalline silicon solar cells *Sol. Energy Mater. Sol. Cells* **147** 115–26

[52] Arruti O A, Gnocchi L, Jeangros Q, Virtuani A and Ballif C 2021 Potential Induced Degradation Mechanism in Rear-Emitter Bifacial Silicon Heterojunction Solar Cells Encapsulated in Different Module Structures *Conference Record of the IEEE Photovoltaic Specialists Conference* (Institute of Electrical and Electronics Engineers Inc.) pp 2032–6

[53] Lee Y and Tay A A O 2013 Stress analysis of silicon wafer-based photovoltaic modules under IEC 61215 mechanical load test *Energy Procedia* vol 33 (Elsevier Ltd) pp 265–71

[54] Sander M, Dietrich S, Pander M, Ebert M and Bagdahn J 2013 Systematic investigation of cracks in encapsulated solar cells after mechanical loading *Sol. Energy Mater. Sol. Cells* **111** 82–9

[55] Pingel S, Frank O, Winkler M, Daryan S, Geipel T, Hoehne H and Berghold J 2010 Potential Induced Degradation of solar cells and panels *2010 35th IEEE Photovoltaic Specialists Conference* (IEEE) pp 002817–22

[56] Crabb R L 1973 Photon Induced Degradation of Electron and Proton Irradiated Silicon Solar Cells *IEEE Trans. Nucl. Sci.* **20** 243–9

[57] Graff K and Pieper H 1975 Carrier lifetime measurements on electron-irradiated silicon crystals *Phys. Status Solidi* **30** 593–9

[58] Schmidt J and Bothe K 2004 Structure and transformation of the metastable boron- and oxygen-related defect center in crystalline silicon *Phys. Rev. B* **69** 024107

[59] Herguth A, Schubert G, Kaes M and Hahn G 2008 Investigations on the long time behavior of the metastable boron–oxygen complex in crystalline silicon *Prog. Photovoltaics Res. Appl.* **16** 135–40

[60] Grant N E, Scowcroft J R, Pointon A I, Al-Amin M, Altermatt P P and Murphy J D 2020 Lifetime instabilities in gallium doped monocrystalline PERC silicon solar cells *Sol. Energy Mater. Sol. Cells* **206**

[61] Grant N E, Altermatt P P, Niewelt T, Post R, Kwapil W, Schubert M C and Murphy J D 2021 Gallium-Doped Silicon for High-Efficiency Commercial Passivated Emitter and Rear Solar Cells *Sol. RRL* **5** 2000754

[62] Sze S M and Ng K K N Physics of Semiconductor Devices, 3rd Edition *Physics of Semiconductor Devices, 3rd Edition* (Hoboken, New Jersey: Wiley Interscience)

[63] Shah A 2020 *Solar Cells and Modules* vol 301 (Cham: Springer International Publishing)

[64] Pander M, Turek M, Bauer J, Luka T, Hagendorf C, Ebert M and Gottschalg R 2018 Benchmarking Light and Elevated Temperature Induced Degradation (LeTID) *35th European Photovoltaic Solar Energy Conference and Exhibition* (Bruxelles)

[65] Ohdaira K, Komatsu Y, Suzuki T, Yamaguchi S and Masuda A 2019 Influence of sodium on the potential-induced degradation for n-type crystalline silicon photovoltaic modules *Appl. Phys. Express* **12** 064004

[66] Dong N C, Islam M A, Ishikawa Y and Uraoka Y 2018 The influence of sodium ions decorated micro-cracks on the evolution of potential induced degradation in p-type crystalline silicon solar cells *Sol. Energy* **174** 1–6

[67] López-Escalante M C, Caballero L J, Martín F, Gabás M, Cuevas A and Ramos-Barrado J R 2016 Polyolefin as PID-resistant encapsulant material in PV modules *Sol. Energy Mater. Sol. Cells* **144** 691–9

[68] Gu W, Li S, Liu X, Chen Z, Zhang X and Ma T 2021 Experimental investigation of the bifacial photovoltaic module under real conditions *Renew. Energy* **173** 1111–22

[69] Lin Y H, Li L Y, Chen S Y, Hsu S P, Chen Y S, Cheng T H, Chen C P and Yu P 2019 New designed electrode patterns on rear side of bifacial PERC solar cells *Conf. Rec. IEEE Photovolt. Spec. Conf.* 1124–6

[70] Dullweber T, Kranz C, Peibst R, Baumann U, Hannebauer H, Fülle A, Steckemetz S, Weber T, Kutzer M, Müller M, Fischer G, Palinginis P and Neuhaus H 2016 PERC+: industrial PERC solar cells with rear Al grid enabling bifaciality and reduced Al paste consumption *Prog. Photovoltaics Res. Appl.* **24** 1487–98

[71] Levrat J, Allebé C, Badel N, Barraud L, Bonnet-Eymard M, Champliaud J, Debrot F, Descoeudres A, Faes A, Lachowicz A, Nicolay S, Sansonnens L, Ballif C, Geissbühler J, De Wolf S and Despeisse M 2014 High-performance hetero-junction crystalline silicon

photovoltaic technology *2014 IEEE 40th Photovoltaic Specialist Conference, PVSC 2014* (Institute of Electrical and Electronics Engineers Inc.) pp 1218–22
[72] Yoshikawa K, Kawasaki H, Yoshida W, Irie T, Konishi K, Nakano K, Uto T, Adachi D, Kanematsu M, Uzu H and Yamamoto K 2017 Silicon heterojunction solar cell with interdigitated back contacts for a photoconversion efficiency over 26% *Nat. Energy* **2**
[73] Dingemans G and Kessels W M M 2012 Status and prospects of Al 2 O 3 -based surface passivation schemes for silicon solar cells *J. Vac. Sci. Technol. A Vacuum, Surfaces, Film.* **30** 040802
[74] Aberle A G 2001 Overview on SiN surface passivation of crystalline silicon solar cells *Sol. Energy Mater. Sol. Cells* **65** 239–48
[75] Glunz S W and Feldmann F 2018 SiO2 surface passivation layers – a key technology for silicon solar cells *Sol. Energy Mater. Sol. Cells* **185** 260–9
[76] Schmidt J, Peibst R and Brendel R 2018 Surface passivation of crystalline silicon solar cells: Present and future *Sol. Energy Mater. Sol. Cells* **187** 39–54
[77] Taguchi M 2021 Review—Development History of High Efficiency Silicon Heterojunction Solar Cell: From Discovery to Practical Use *ECS J. Solid State Sci. Technol.* **10** 025002
[78] Serenelli L, Martini L, Menchini F, Izzi M, de Cesare G, Condorelli G, Gerardi C, Muñoz D and Tucci M 2021 Selective contacts and fill factor limitations in heterojunction solar cells *Prog. Photovoltaics Res. Appl.* 1–9
[79] Balent J, Smole F, Topic M and Krc J 2021 Numerical analysis of selective ITO/a-Si:H contacts in heterojunction silicon solar cells: Effect of Defect states in doped a-Si:H Layers on Performance Parameters *IEEE J. Photovoltaics* **11** 634–47
[80] Condorelli G, Favre W, Battaglia A, Rotoli P, Canino A, Sciuto M, Ragonesi A, Danel A, Munoz D, Roux C, Lerat J F, Medlege F, Barth V, Sicot L, Ribeyron P J and Gerardi C 2018 High Efficiency Hetero-Junction: From Pilot Line to Industrial Production *2018 IEEE 7th World Conference on Photovoltaic Energy Conversion, WCPEC 2018 - A Joint Conference of 45th IEEE PVSC, 28th PVSEC and 34th EU PVSEC* (Institute of Electrical and Electronics Engineers Inc.) pp 1970–3
[81] Condorelli G, Rotoli P, Canino A, Battaglia A, Favre W, Ozanne A-S, Moustafa A, Danel A, Munoz D, Ribeyron P-J and Gerardi C 2017 Contamination control challenges on SHJ solar cell processing *2017 IEEE 44th Photovoltaic Specialist Conference (PVSC)* (IEEE) pp 1747–51
[82] Izzi M, Tucci M, Veneri P D, Scalari S, Proietti D, Colletti C, Balucani M and Serenelli L 2018 AMPERE: An European project aimed to decrease the Levelized Cost of Energy with innovative heterojunction bifacial module solution ready for the market. *2018 IEEE 7th World Conference on Photovoltaic Energy Conversion (WCPEC) (A Joint Conference of 45th IEEE PVSC, 28th PVSEC & 34th EU PVSEC)* (IEEE) pp 569–72
[83] Madumelu C, Wright B, Soeriyadi A, Wright M, Chen D, Hoex B and Hallam B 2020 Investigation of light-induced degradation in N-Type silicon heterojunction solar cells during illuminated annealing at elevated temperatures *Sol. Energy Mater. Sol. Cells* **218**
[84] Cattin J, Senaud L Lou, Haschke J, Paviet-Salomon B, Despeisse M, Ballif C and Boccard M 2021 Influence of light soaking on silicon heterojunction solar cells with various architectures *IEEE J. Photovoltaics* **11** 575–83
[85] Haschke J, Dupré O, Boccard M and Ballif C 2018 Silicon heterojunction solar cells: Recent technological development and practical aspects - from lab to industry *Sol. Energy Mater. Sol. Cells* **187** 140–53
[86] Chen S, Manders J R, Tsang S W and So F 2012 Metal oxides for interface engineering in polymer solar cells *J. Mater. Chem.* **22** 24202–12
[87] Greiner M T and Lu Z H 2013 Thin-film metal oxides in organic semiconductor devices: Their electronic structures, work functions and interfaces *NPG Asia Mater.* **5** 1–16
[88] Il Park S, Jae Baik S, Im J S, Fang L, Jeon J W and Su Lim K 2011 Towards a high efficiency amorphous silicon solar cell using molybdenum oxide as a window layer instead of conventional p-type amorphous silicon carbide *Appl. Phys. Lett.* **99**
[89] Fang L, Baik S J, Kim J W, Kang S J, Seo J W, Jeon J-W, Kim Y H and Lim K S 2011 Tunable work function of a WO x buffer layer for enhanced photocarrier collection of pin - type amorphous silicon solar cells *J. Appl. Phys.* **109** 104501
[90] Battaglia C, Yin X, Zheng M, Sharp I D, Chen T, McDonnell S, Azcatl A, Carraro C, Ma B, Maboudian R, Wallace R M and Javey A 2014 Hole Selective MoO x Contact for Silicon Solar Cells *Nano Lett.* **14** 967–71

[91] Bivour M, Temmler J, Steinkemper H and Hermle M 2015 Molybdenum and tungsten oxide: High work function wide band gap contact materials for hole selective contacts of silicon solar cells *Sol. Energy Mater. Sol. Cells* **142** 34–41
[92] Dreon J, Cattin J, Christmann G, Febba D, Paratte V, Antognini L, Lin W, Nicolay S, Ballif C and Boccard M 2021 Performance Limitations and Analysis of Silicon Heterojunction Solar Cells Using Ultra-Thin MoOxHole-Selective Contacts *IEEE J. Photovoltaics* **11** 1158–66
[93] Shewchun J, Burk D and Spitzer M B 1980 MIS and SIS solar cells *IEEE Trans. Electron Devices* **27** 705–16
[94] Geissbühler J, Werner J, Martin De Nicolas S, Barraud L, Hessler-Wyser A, Despeisse M, Nicolay S, Tomasi A, Niesen B, De Wolf S and Ballif C 2015 22.5% Efficient Silicon Heterojunction Solar Cell With Molybdenum Oxide Hole Collector *Appl. Phys. Lett.* **107**
[95] Dréon J, Jeangros Q, Cattin J, Haschke J, Antognini L, Ballif C and Boccard M 2020 23.5%-Efficient Silicon Heterojunction Silicon Solar Cell Using Molybdenum Oxide As Hole-Selective Contact *Nano Energy* **70**
[96] Essig S, Dréon J, Rucavado E, Mews M, Koida T, Boccard M, Werner J, Geissbühler J, Löper P, Morales-Masis M, Korte L, De Wolf S and Ballif C 2018 Toward Annealing-Stable Molybdenum-Oxide-Based Hole-Selective Contacts For Silicon Photovoltaics *Sol. RRL* **2** 1–5
[97] Macco B, Vos M F J, Thissen N F W, Bol A A and Kessels W M M 2015 Low-temperature atomic layer deposition of MoOx for silicon heterojunction solar cells *Phys. Status Solidi - Rapid Res. Lett.* **9** 393–6
[98] Bivour M, Macco B, Temmler J, Kessels W M M and Hermle M 2016 Atomic Layer Deposited Molybdenum Oxide for the Hole-selective Contact of Silicon Solar Cells *Energy Procedia* **92** 443–9
[99] Vos M F J, Macco B, Thissen N F W, Bol A A and Kessels W M M (Erwin) 2016 Atomic layer deposition of molybdenum oxide from (N t Bu) 2 (NMe 2) 2 Mo and O 2 plasma *J. Vac. Sci. Technol. A Vacuum, Surfaces, Film.* **34** 01A103
[100] Ziegler J, Mews M, Kaufmann K, Schneider T, Sprafke A N, Korte L and Wehrspohn R B 2015 Plasma-enhanced atomic-layer-deposited MoO x emitters for silicon heterojunction solar cells *Appl. Phys. A* **120** 811–6
[101] Poodt P, Cameron D C, Dickey E, George S M, Kuznetsov V, Parsons G N, Roozeboom F, Sundaram G and Vermeer A 2012 Spatial atomic layer deposition: A route towards further industrialization of atomic layer deposition *J. Vac. Sci. Technol. A Vacuum, Surfaces, Film.* **30** 010802
[102] Tyagi A, Biswas J, Ghosh K, Kottantharayil A and Lodha S 2021 Performance Analysis of Silicon Carrier Selective Contact Solar Cells With ALD MoOx as Hole Selective Layer *Silicon*
[103] Yang X, Zheng P, Bi Q and Weber K 2016 Silicon heterojunction solar cells with electron selective TiOx contact *Sol. Energy Mater. Sol. Cells* **150** 32–8
[104] Wang W, He J, Yan D, Samundsett C, Phang S P, Huang Z, Shen W, Bullock J and Wan Y 2020 21.3%-efficient n-type silicon solar cell with a full area rear TiOx/LiF/Al electron-selective contact *Sol. Energy Mater. Sol. Cells* **206** 110291
[105] Martí A and Araújo G L 1996 Limiting efficiencies for photovoltaic energy conversion in multigap systems *Sol. Energy Mater. Sol. Cells* **43** 203–22
[106] Sasaki K, Agui T, Nakaido K, Takahashi N, Onitsuka R and Takamoto T 2013 Development of InGaP/GaAs/InGaAs inverted triple junction concentrator solar cells pp 22–5
[107] Aydin E, Allen T G, De Bastiani M, Xu L, Ávila J, Salvador M, Van Kerschaver E and De Wolf S 2020 Interplay between temperature and bandgap energies on the outdoor performance of perovskite/silicon tandem solar cells *Nat. Energy* **5** 851–9
[108] De Bastiani M, Subbiah A S, Aydin E, Isikgor F H, Allen T G and De Wolf S 2020 Recombination junctions for efficient monolithic perovskite-based tandem solar cells: Physical principles, properties, processing and prospects *Mater. Horizons* **7** 2791–809
[109] Al-Ashouri A, Köhnen E, Li B, Magomedov A, Hempel H, Caprioglio P, Márquez J A, Vilches A B M, Kasparavicius E, Smith J A, Phung N, Menzel D, Grischek M, Kegelmann L, Skroblin D, Gollwitzer C, Malinauskas T, Jošt M, Matič G, Rech B, Schlatmann R, Topič M, Korte L, Abate A, Stannowski B, Neher D, Stolterfoht M, Unold T, Getautis V and Albrecht S 2020 Monolithic perovskite/silicon tandem solar cell with >29% efficiency by enhanced hole extraction *Science (80-.).* **370** 1300–9

[110] Werner J, Niesen B and Ballif C 2018 Perovskite/Silicon Tandem Solar Cells: Marriage of Convenience or True Love Story? – An Overview *Adv. Mater. Interfaces* **5** 1–19
[111] Kimura T, Goto T, Shintani H, Ishizaka K, Arima T and Tokura Y 2003 Magnetic control of ferroelectric polarization *Nature* **426** 55–8
[112] Bednorz J G and Müller K A 1988 Perovskite-Type Oxides—the New Approach to High-Tc Superconductivity. Nobel Lecture *Angew. Chemie Int. Ed. English* **27** 735–48
[113] Im J-H, Lee C-R, Lee J-W, Park S-W and Park N-G 2011 6.5% efficient perovskite quantum-dot-sensitized solar cell *Nanoscale* **3** 4088
[114] A K, K T, Y S and T M 2009 Organometal halide perovskites as visible-light sensitizers for photovoltaic cells *J. Am. Chem. Soc.* **131** 6050–1
[115] Kim H S, Lee C R, Im J H, Lee K B, Moehl T, Marchioro A, Moon S J, Humphry-Baker R, Yum J H, Moser J E, Grätzel M and Park N G 2012 Lead iodide perovskite sensitized all-solid-state submicron thin film mesoscopic solar cell with efficiency exceeding 9% *Sci. Rep.* **2**
[116] Xu J, Boyd C C, Yu Z J, Palmstrom A F, Witter D J, Larson B W, France R M, Werner J, Harvey S P, Wolf E J, Weigand W, Manzoor S, Van Hest M F A M, Berry J J, Luther J M, Holman Z C and McGehee M D 2020 Triple-halide wide-band gap perovskites with suppressed phase segregation for efficient tandems *Science (80-.).* **367** 1097–104
[117] Noh J H, Im S H, Heo J H, Mandal T N and Seok S Il 2013 Chemical management for colorful, efficient, and stable inorganic-organic hybrid nanostructured solar cells *Nano Lett.* **13** 1764–9
[118] Sadhanala A, Ahmad S, Zhao B, Giesbrecht N, Pearce P M, Deschler F, Hoye R L Z, Gödel K C, Bein T, Docampo P, Dutton S E, De Volder M F L and Friend R H 2015 Blue-Green Color Tunable Solution Processable Organolead Chloride–Bromide Mixed Halide Perovskites for Optoelectronic Applications *Nano Lett.* **15** 6095–101
[119] Saliba M, Matsui T, Seo J, Domanski K, Correa-Baena J-P, Nazeeruddin M K, Zakeeruddin S M, Tress W, Abate A, Hagfeldt A and Grätzel M 2016 Cesium-containing triple cation perovskite solar cells: - SUPPORTING *Energy Environ. Sci.* **9** 1989–97
[120] Gil-Escrig L, Momblona C, La-Placa M G, Boix P P, Sessolo M and Bolink H J 2018 Vacuum Deposited Triple-Cation Mixed-Halide Perovskite Solar Cells *Adv. Energy Mater.* **8**
[121] Mannino G, Deretzis I, Smecca E, La Magna A, Alberti A, Ceratti D and Cahen D 2020 Temperature-Dependent Optical Band Gap in CsPbBr3, MAPbBr3, and FAPbBr3 Single Crystals *J. Phys. Chem. Lett.* **11** 2490–6
[122] Abdollahi Nejand B, Hossain I M, Jakoby M, Moghadamzadeh S, Abzieher T, Gharibzadeh S, Schwenzer J A, Nazari P, Schackmar F, Hauschild D, Weinhardt L, Lemmer U, Richards B S, Howard I A and Paetzold U W 2020 Vacuum-Assisted Growth of Low-Bandgap Thin Films (FA0.8MA0.2Sn0.5Pb0.5I3) for All-Perovskite Tandem Solar Cells *Adv. Energy Mater.* **10**
[123] Smecca E, Numata Y, Deretzis I, Pellegrino G, Boninelli S, Miyasaka T, La Magna A and Alberti A 2016 Stability of solution-processed MAPbI3 and FAPbI3 layers *Phys. Chem. Chem. Phys.* **18** 13413–22
[124] Deretzis I, Smecca E, Mannino G, La Magna A, Miyasaka T and Alberti A 2018 Stability and Degradation in Hybrid Perovskites: Is the Glass Half-Empty or Half-Full? *J. Phys. Chem. Lett.* **9** 3000–7
[125] Saliba M, Matsui T, Seo J Y, Domanski K, Correa-Baena J P, Nazeeruddin M K, Zakeeruddin S M, Tress W, Abate A, Hagfeldt A and Grätzel M 2016 Cesium-containing triple cation perovskite solar cells: Improved stability, reproducibility and high efficiency *Energy Environ. Sci.* **9** 1989–97
[126] Sutter-Fella C M, Ngo Q P, Cefarin N, Gardner K L, Tamura N, Stan C V., Drisdell W S, Javey A, Toma F M and Sharp I D 2018 Cation-Dependent Light-Induced Halide Demixing in Hybrid Organic-Inorganic Perovskites *Nano Lett.* **18** 3473–80
[127] Wang P, Wu Y, Cai B, Ma Q, Zheng X and Zhang W H 2019 Solution-Processable Perovskite Solar Cells toward Commercialization: Progress and Challenges *Adv. Funct. Mater.* **29** 1–37
[128] Dewi H A, Li J, Wang H, Chaudhary B, Mathews N, Mhaisalkar S and Bruno A 2021 Excellent Intrinsic Long-Term Thermal Stability of Co-Evaporated MAPbI3 Solar Cells at 85 °C *Adv. Funct. Mater.* **31**
[129] Li J, Wang H, Chin X Y, Dewi H A, Vergeer K, Goh T W, Lim J W M, Lew J H, Loh K P, Soci C, Sum T C, Bolink H J, Mathews N, Mhaisalkar S and Bruno A 2020 Highly Efficient

Thermally Co-evaporated Perovskite Solar Cells and Mini-modules *Joule* **4** 1035–53
[130] Shi L, Bucknall M P, Young T L, Zhang M, Hu L, Bing J, Lee D S, Kim J, Wu T, Takamure N, McKenzie D R, Huang S, Green M A and Ho-Baillie A W Y 2020 Gas chromatography-mass spectrometry analyses of encapsulated stable perovskite solar cells *Science (80-.).* **368**
[131] Liu J, Aydin E, Yin J, De Bastiani M, Isikgor F H, Rehman A U, Yengel E, Ugur E, Harrison G T, Wang M, Gao Y, Khan J I, Babics M, Allen T G, Subbiah A S, Zhu K, Zheng X, Yan W, Xu F, Salvador M F, Bakr O M, Anthopoulos T D, Lanza M, Mohammed O F, Laquai F and De Wolf S 2021 28.2%-efficient, outdoor-stable perovskite/silicon tandem solar cell *Joule* **5** 3169–86
[132] Chen B, Yu Z J, Manzoor S, Wang S, Weigand W, Yu Z, Yang G, Ni Z, Dai X, Holman Z C and Huang J 2020 Blade-Coated Perovskites on Textured Silicon for 26%-Efficient Monolithic Perovskite/Silicon Tandem Solar Cells *Joule* **4** 850–64
[133] Sahli F, Werner J, Kamino B A, Bräuninger M, Monnard R, Paviet-Salomon B, Barraud L, Ding L, Diaz Leon J J, Sacchetto D, Cattaneo G, Despeisse M, Boccard M, Nicolay S, Jeangros Q, Niesen B and Ballif C 2018 Fully textured monolithic perovskite/silicon tandem solar cells with 25.2% power conversion efficiency *Nat. Mater.* **17** 820–6
[134] Hou Y, Aydin E, De Bastiani M, Xiao C, Isikgor F H, Xue D-J, Chen B, Chen H, Bahrami B, Chowdhury A H, Johnston A, Baek S-W, Huang Z, Wei M, Dong Y, Troughton J, Jalmood R, Mirabelli A J, Allen T G, Van Kerschaver E, Saidaminov M I, Baran D, Qiao Q, Zhu K, De Wolf S and Sargent E H 2020 Efficient tandem solar cells with solution-processed perovskite on textured crystalline silicon *Science (80-.).* **367** 1135–40
[135] Lu H, Liu Y, Ahlawat P, Mishra A, Tress W R, Eickemeyer F T, Yang Y, Fu F, Wang Z, Avalos C E, Carlsen B I, Agarwalla A, Zhang X, Li X, Zhan Y, Zakeeruddin S M, Emsley L, Rothlisberger U, Zheng L, Hagfeldt A and Grätzel M 2020 Vapor-assisted deposition of highly efficient, stable black-phase FAPbI3 perovskite solar cells *Science (80-.).* **370**
[136] Lamanna E, Matteocci F, Calabrò E, Serenelli L, Salza E, Martini L, Menchini F, Izzi M, Agresti A, Pescetelli S, Bellani S, Del Río Castillo A E, Bonaccorso F, Tucci M and Di Carlo A 2020 Mechanically Stacked, Two-Terminal Graphene-Based Perovskite/Silicon Tandem Solar Cell with Efficiency over 26% *Joule* **4** 865–81

Design of nanostructured coating to prevent biofilm formation on surfaces

Daniele Marra[1], Federica Recupido[1], Angela Di Somma[2], Carolina Canè[2], Annalisa Acquesta[1], Giuseppe Toscano[1], Tullio Monetta[1], Angela Duilio[2], Sergio Caserta[1,3].

[1] DICMaPI, Università di Napoli Federico II, P.le V.Tecchio 80, 80125 Napoli, Italy.
[2] Department of Chemical Sciences, University of Naples "Federico II" Via Cinthia 4, 80126 Napoli, Italy.
[3] CEINGE, Advanced Biotechnologies, 80145, Naples, Italy.

E-mail: daniele.marra2@unina.it; sergio.caserta@unina.it

Abstract. Biofilms are surface-attached microbial organizations, where microbial species are enclosed in an exo-polysaccharides matrix. Its removal/prevention from surfaces represents an important challenge and a relevant issue in many fields. In this light, antimicrobial peptides (AMPs) represent good candidates to conventional antibiotics due to their selectivity towards targets (cell membranes), rapid action and low tendency to induce bacterial resistance. In our project we are investigating novel techniques to realise nanostructured coatings on surfaces, in order to prevent bacterial adhesion and biofilm formation. We are focusing our attention on a well-known antimicrobial peptide (Magainin-2). Two bacteria models have been considered: *E.coli* and *P. fluorescens*. Our goal is to optimize AMPs grafting on surfaces by means of treatment methodologies based on Oxygen plasma technologies. In this work we present a methodology to investigate anti-biofilm efficiency of the nano-functionalized surfaces, compared with the case of the raw surfaces. Biofilm morphologies can be measured using Confocal Laser Scanning Microscopy and image analysis techniques to quantify its structure by measuring quantitative morphological parameters.

1. Introduction

Microorganisms are ubiquitous in almost all environments due their resilience and adaptability able to protect themselves from external agents. Bacteria can be found in planktonic form or in specific conditions, as sessile aggregates on both biotic and abiotic surfaces originating complex structures known as biofilm [1].

Biofilm are structured communities of microorganisms enclosed in an extracellular polymer matrix (EPS), which is produced by bacteria themselves and consists of polysaccharides, proteins and extracellular DNA [2]. Surfaces adhesion, and EPS production provide a complex physical barrier able to overcome many external stresses that can be normally dangerous for planktonic microorganism. This imposed stimuli include physical stresses such as mechanical (e.g. flow induced [3]), thermal (e.g. temperature variation), and dehydration [4-6]; and (bio-)chemical stresses such as the ones related to the presence of sanitizer/cleaning solutions, biological agents (e.g. antibiotics [7]). The mechanism involved in chemical resistance of biofilms is related to its tridimensional porous structure which hinders the diffusional penetration of antimicrobials, and other bactericidal substances [8].

Biofilms represent a biological system with a high degree of organization, it is a viscoelastic gel in which bacteria are suspended and adhered and coordinated in communities able to cooperate in metabolic-reproductive and infectious processes. Biofilms create at the interface between two different phases, typically at the liquid-air interface (airborne biofilms) or at solid-liquid interface (submerged biofilms). We are mainly focusing here on the second type of biofilm, that are particularly relevant in the case of biocontamination and biofouling of surfaces. For this reason, the biofilm is an ubiquitous structure, and due to its nature is an issue in several industrial and health-related applications, such as contamination of medical equipment (e.g. central venous, urinary catheters, heart valves, intrauterine devices, dental implants [9]), bio-corrosion of industrial equipment, biofouling [10], and cross-contamination of industrial products [7]. A part for biosafety, the most problematic issue is the demanding and expansive removal processes [11]. On the other hand, there are also positive important applications of biofilms mainly related to their implementation in many biotechnological processes such as bioremediation [12], wastewater treatment [13], synthesis of fine chemicals [14] and biofuels production [15].

Once attached on a solid surface, the outer surface of biofilm (facing the submerging fluid) is constantly evolving, and it results in the formation of a chemical-physical gradient that influence also the adhesion of the biofilm on the surfaces. In this contest the development of surfaces coated with antibiofilm substances could represent a truly effective overcoming of the problem related to biofilm. Novel anti-biofilm compounds are represented by the class of AntiMicrobialPeptides (AMPs), a class of effector molecules of the innate immune system with a broad antimicrobial spectrum [16]. In addition to direct antimicrobial function, AMPs can exploit anti-biofilm activity by targeting bacterial proteins involved in the biofilm formation [17, 18] or by inhibiting the adhesion of bacterial cells on solid surfaces.

Due to the great variety of amino acid sequences and the consequent AMPs structural features, the exact action mechanism of action of AMPs is still a controversial issue. However, a commonly accepted interpretation is related to the positive charge of AMPs that induce electrostatic interactions with negatively charged microbial membranes through bivalent cation exchange [19].

The use of AMPs in adsorption processes, functionalization, and immobilization of organic coatings onto substrate surfaces was object of a growing interest in the literature, in particular in the field of clinical applications [20, 21]. Recent studies demonstrated that AMP coupled with different substrates have an effective action against biofilm formation [22]. Moreover, AMPs coatings have several properties capable of overcoming the negative aspects of other antibacterial coatings, such as the risk of developing bacterial resistance, short-term antimicrobial protection, limited antimicrobial spectrum and high cytotoxicity [22].

A promising and innovative method for covalent bonding and/or AMPs immobilization on different types of surfaces is represented by cold plasma technology.

Cold, or non-thermal, plasmas are a partially or fully ionized gas. Plasmas are macroscopically neutral, but contain photons, electrons, positive and negative ions, free radicals, reaction products among them, as well as atoms and molecules, which can be in different states of excitation depending on the energy state of the system. Cold plasma treatments can be used to modify any material (polymeric, metal or ceramic), as the average temperature of the gases in contact with the material is similar to the ambient one. Since the 1970s, cold plasma-based technologies have increasingly been established in the industrial field to realise coatings, surfaces activations, and modifications. Depending on the applications, plasma technologies are split into two main categories: those involving the deposition of an organic or inorganic film (plasma coating) and those that use the plasma to activate/modify the surface with or without its functionalization (plasma treatment) [23]. Some kinds of plasma may have a general oxidative effect on cells, inhibiting their growth. Antibacterial activity of cold plasmas is industrially interesting because it is non-pollutant and may not require an expensive setup [24].

In this work an innovative approach to inhibit bio-contamination and evaluate antibiofilm activity of AMPs functionalized surfaces is proposed.

2. Materials and methods

2.1. *Microorganism and culture conditions*

A collection of EUCAST reference strains was used in this work. Gram-negative bacteria *Escherichia coli* (ATCC 25922), *Escherichia coli* TEM-1 β-lactamase (ATCC 35218), *Pseudomonas aeruginosa* (ATCC 27853), *Klebsiella pneumoniae* (ATCC 700603) were cultured on Luria Bertani agar and broth; *Haemophilus influenzae* (NCTC 12975) was cultured on Chocolate agar plate and Luria Bertani broth. Gram-positive methicillin-resistant bacteria *Staphylococcus aureus* (MRSA) (NCTC 12493) and *Enterococcus faecalis* (NTCC 12697) were cultured respectively on Luria Bertani agar and broth while *Streptococcus pneumonia* (NCTC 12977) on Blood plate and in BHI broth.

All bacteria were incubated aerobically at 37°C for 18 h except *Streptococcus pneumonia*, which was incubated in CO_2-enriched atmosphere.

Pseudomonas fluorescens AR 11, provided by DSMZ collection, was growth in a sterile M9-minimal medium (6.8 g/L of Na_2HPO_4, 3 g/L of KH_2PO_4, 0.5 g/L of NaCl, and 1 g/L NH_4Cl, 0.24 g/L $MgSO_4{\cdot}7H_2O$, 0.04 g/L $CaCl_2{\cdot}2H_2O$, 0.05 g/L EDTA, 8.3 mg/L $FeCl_3$, 0.84 mg/L $ZnCl_2$, 0.1 mg/L $CuCl_2{\cdot}2\,H_2O$, 0.1 mg/L $CoCl_2{\cdot}2H_2O$, 0.1 mg/L H_3BO_3, and 0.016 mg/L $MnCl_2{\cdot}4\,H_2O$ supplemented with 0.4% succinic acid, at pH 7). The obtained suspension was then incubated at 30 °C in a shaken vessel (90 rpm), to achieve an optical density (OD_{600nm}) ≈ 0.5 cm^{-1}, which corresponds to 10^7 cells/mL, approximately.

2.2. *Bacterial susceptibility testing*

The minimum inhibitory concentration (MIC) of Mag-2 was calculated by broth microdilution. The bacterial strains were incubated overnight in LB at 37°C. The cultures were diluted approximately to 5 $X10^5$ CFU/mL and 50 µL of bacterial suspension was added to ten wells and incubated with serial dilutions of peptide from an initial concentration of 512 µM. The sterility control well contained 100 µL of LB, while the growth control well contained 100 µL of microbial suspension. The plates were incubated at 37°C for 16/ 20h of grown and the MIC was determined by the lowest concentration showing no visible growth by measuring the absorbance at 600 nm. For the minimum bactericidal concentration (MBC), the MIC well and the wells with concentration higher than MIC value were plated on agar plate and incubated at 37°C for 16/ 20h. The MBC value was determined as the amount of AMPs allowing no colony growth from the directly plated content of the wells. All MIC and MBC determinations of all assays were performed in triplicate.

2.3. *Biofilm formation under Mag-2 treatment*

Preliminary tests on Mag-2 were performed under stagnant conditions (Fig. 2a), Mag-2 treated biofilms were formed in 12 well-microplates at 37°C. In each well the bacterial strain of *E. coli* was inoculated at a concentration of 0.5 OD/mL with LB medium and 1% glucose. The multiwells were incubated in aerobiosis for 24 hours at 37°C in the presence and absence of different concentration of Mag-2. To assess the antibiofilm properties of Mag-2, Crystal Violet assay was used. The liquid medium was removed and the wells washed with sterile distilled water. The plates were then stained with Crystal Violet 0.1% for 15 min. Excess stain was rinsed off by placing the plate under running tap water. After this the plates were air dried, the dye bound to the adherent cells was solubilized with 20% (v/v) glacial acetic acid and 80% (v/v) ethanol per well. The OD of each well was measured at 590 nm.

2.4. *Oxygen plasma samples functionalization*

The implementation of Cold Plasma surface treatments are obtained by a preliminary experimental campaign aimed to provide the best operating condition for AMPs-substrate couple. Different plasma parameters can be changed in order to obtain the most suitable immobilization of AMPs on the sample's surface. In particular, the gas pressure, the treatment duration and the electrical power. Infact, they play an important role in the effectiveness of the samples surface modification. To do this, optical clear coupons slides, made of Polymethylmethacrylate (PMMA) (1.8 × 1.8 cm), are treated by placing them into cold plasma reactor fed with pure oxygen. After plasma treatment, samples are dipped in the AMPs aqueous solution to induce the surface immobilization of the peptide. Untreated samples are used as a control to evaluate the antibiofilm activity of treated ones. Moreover, the change in samples surface energy and in the chemical surface composition are used to determine the surface treatment effect.

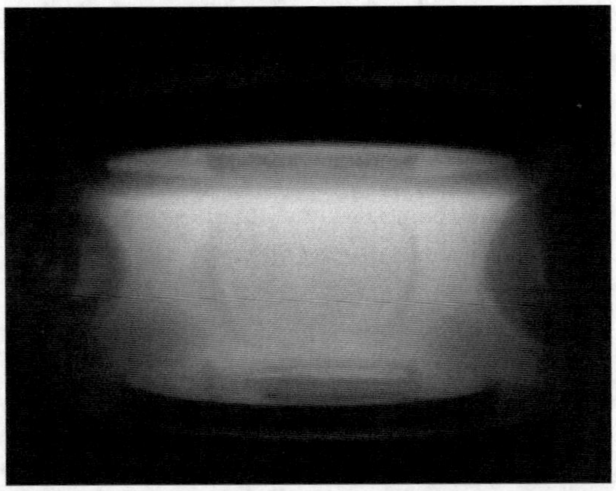

Figure 1. Inside view of a coldplasma reaction chamber fed by using oxygen. The figure clearly shows the horizontal setup of the two electrodes (top and bottom of the figure) the yellow glow and the dark shield.

2.5. Biofilm growth conditions

We considered different possible biofilm growth method. In particular we cultivated biofilms under different flow conditions, following the protocols validated by Recupido et al 2020 [16]. This method can be applied for different aerobic bacterial strain and for different types of coatings. In this work, *E.coli* biofilm was growth in static condition with different concentration of Mag-2 while *P. fluorescens* biofilm was growth under all three different conditions: two batch techniques, (stagnant fluid and shaken vessels) and a microfluidic technique based on a custom-made continuous flow apparatus. Different flow conditions can significantly affect transport phenomena efficiency and biocontamination mechanisms, for this reason, an efficient testing must consider the influence of flow. Experiments are carried out for an overall incubation time of 72 h. The three experimental set-ups are shown in Fig. 2. In all conditions the set-up allow to host either coated and or uncoated coupons substrates with the goal to evaluate the effect of surface nanostructured treatments on biofilm formation and growth.

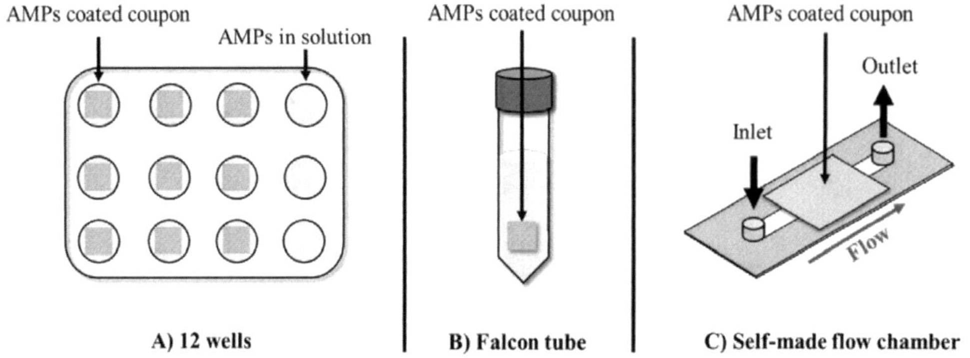

Figure 2. Schematization of three set-up proposed for biofilm growth. Stagnant fluid (A), shaken vessel (B) and in flow setup (C) are designed to host functionalized coupons to evaluate antibiofilm coating at different flow conditions.

2.5.1. No flow. Biofilms are cultivated over optical clear coupons slides (1.8 × 1.8 cm) at 30 °C for *P. fluorescens* and 37°C for *E.coli*, placed inside 12 aerated well-microplates. A coupon of chosen material (e.g.: glass, polycarbonate, etc.), is inserted in each well and submerged with 1.8 mL of sterile minimal medium (details on composition are reported in paragraph 2.1) and 200 µl of inoculum (1:10 dilution). During incubation, biofilm is formed on both sides of the coupon. In this work we used glass coupons with a total surface area of 6.48 cm^2, the ratio of active surface/total volume in each well was 8.5 cm^{-1}.

2.5.2. Shaken vessels. Biofilms are cultivated over optical clear coupons slides hosted in tilted 50 mL Falcon tubes. Each Falcon tube is filled with 9 mL of sterile minimal medium and 1 mL of inoculum (1:10 dilution) at 0.5 OD. Falcons are placed on an oscillating plate (90 rpm) of a shaking incubator at 30 °C. After 24 h of incubation the turbidity of the medium increases considerably (up to about 1.0 cm^{-1}), implying rapid lack of food and approach of death phase of the bacteria, so we decided to replace the nutrients every 24 h. The experiment is carried out for an overall time of 72h, medium was replaced twice.

2.5.3. Microfluidic setup. We proposed a custom-made flow cell apparatus. Constructive details were reported previously [3]. Flow cell hosts a coupon plate, in this work we used a optical glass slide (26 mm x 76 mm x 1 mm). The apparatus was sterilized using a cleaning solution (70 % ethanol) and autoclave treatment for all the metal components. The not autoclavable components (such as tubes and valves) were sterilized using 2 % solution of Steril C, an anionic surfactant, corresponding of 25 %-30 % (w/w) of peracetic acid, and rinsed with sterile deionized water. The flow apparatus was first preconditioned with fresh medium, and then was filled with 10 mL of bacterial suspension, through a syringe and a three-way valve. Then, the liquid was left stagnant for 2 hours, to allow bacterial cells adhering to surface. After this adhesion time, constant flow (Q) was applied through a peristaltic pump. Such in-flow apparatus enables to impose controlled shear stress conditions by monitoring the inlet flow rate to the microfluidic flow cell. The wall shear rate is a function of the imposed volumetric flow rate and the flow cell geometrical characteristics: $\dot{\gamma}_W = \frac{3\,Q}{(2\,\delta^2 \cdot W)}$. Where δ and W are the half thickness and the width, respectively, of the rectangular section of the microfluidic chamber. Experiments were run for 72 h, which corresponds to the biofilm maturation phase. Reynolds number $Re = \frac{\rho \cdot v \cdot D_h}{\eta}$, calculated considering the hydraulic diameter $D_h = \frac{4A}{P} = \frac{4\,(W \cdot \delta)}{W + 2\delta}$ was evaluated to ensure laminar flow condition. In the experimental investigation reported in this work $\dot{\gamma}_w$ was changed in the range 0.1 - 0.4 - 1 s^{-1}.

2.6. *Biofilm imaging and analysis.*

Biofilm morphology was quantified by analysis of Confocal Scanning Laser Microscopy (CSLM) images. Fluorescent staining was used to visualize cells and EPS on different wavelength. Following the manufacture protocols (ThermoFisher, 2018), cells were stained using the green, fluorescent dye SYTO® 9 (3 µL/mL), while EPS lectins were stained with the red dye Texas Red TRITC conjugate CoA (100 µg/mL) (Tetra-methyl-rhodamine conjugated Concanavalin A, Molecular Probes, Invitrogen, Carlsbad, CA, USA).
Images were acquired by using an inverted microscope (Nikon Eclipse Ti E, Nikon Corporation, Japan), using a Plan Apo λ 40 X/1.49 NA air-objective and a Nikon digital camera. Excitation was provided at the wavelength of 488.6 nm using a detection filter of 498 nm and at the wavelength of 561.5 nm with a detection filter of 580 nm. Z-stacks of samples were acquired along vertical direction using 0.5 µm-interval.

3. Results

3.1. *Susceptibility essay*

Preliminary experiments were carried out to evaluate the antimicrobial activity of Magainin-2 (Mag-2) an AMP derived from the African clawed frog Xenopus laevis. Different EUCAST (European Committee Antimicrobial Susceptibility Testing) bacterial strains were grown in the presence of serial dilution of Mag-2 from 512 µM to 0.5µM and the MIC values were determined as the lowest concentration showing no visible growth after 24h of incubation at 37°C.

In addition to the Minimum Inhibitory Concentration (MIC) values, the bactericidal activity of the AMPs was evaluated by determining the Minimum Bactericidal Concentrations (MBCs). The MIC and MBC values of Mag-2 obtained in these experiments are summarized in Table 1.

Table 1. Minimum inhibitory concentrations (MIC) and minimum bactericidal concentrations (MBC) Mag-2

Strain	MIC (µg/mL)	MBC (µg/mL)
S. aureus NCTC 12493	64	64
E. coli ATCC 35218	64	64
E. coli ATCC 25922	128	64
K.Pneumoniae	> 512	512
P. aeruginosa	> 512	512
S.aureus	> 512	> 512
H. Influenzae	16	64
S. pneumoniae	32	256
E. feacalis	> 512	> 512

3.2. *AMP effect on biofilm morphology*

The in vitro effect of Mag-2 on *E. coli* biofilm formation was first evaluated in static condition (no flow). Bacterial cells were inoculated in the presence and in the absence of three different concentrations of Mag-2, 10, 25 and 50 μM, respectively. Quantification of biofilm formation was monitored by crystal violet assay. Figure 4 shows that Mag-2 decreased biofilm development in a dose-dependent manner displaying a reduction around 80% in biofilm growth at 50 μM Mag-2.

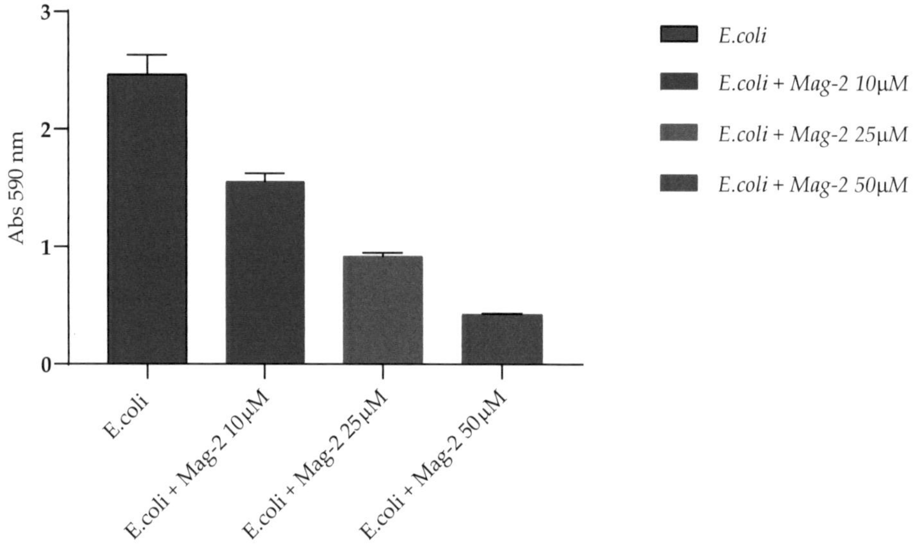

Figure 3. The effect of Mag-2 on the biofilm formation. The bacterial cells were incubated for 24 hours at 37°C in presence and in absence of Mag-2. The error bars on the graphs stand for the standard deviation from the mean of 3 experiments. Oneway ANOVA statistical test using Graphpad prism software was performed, $p < 0.0001$.

3.3. AMP effect on biofilm morphology

In addition, the morphology of *E.coli* biofilm formed on glass substrates was assessed by Confocal Laser Scanning Microscopy (CLSM) Z-stack using a double staining procedure. The case of stagnant fluid is here reported.

In the Fig. 4a the untreated bacteria exhibited a compact and thick biofilm matrix with a complex architecture. In the presence of 25μM and 50 μM of Mag-2, an alteration in biofilm architecture was clearly detected (Fig. 4b, c). In particular, the biofilm thickness was gradually decreased while bacterial cells were clearly indicated by the intense green fluorescence.

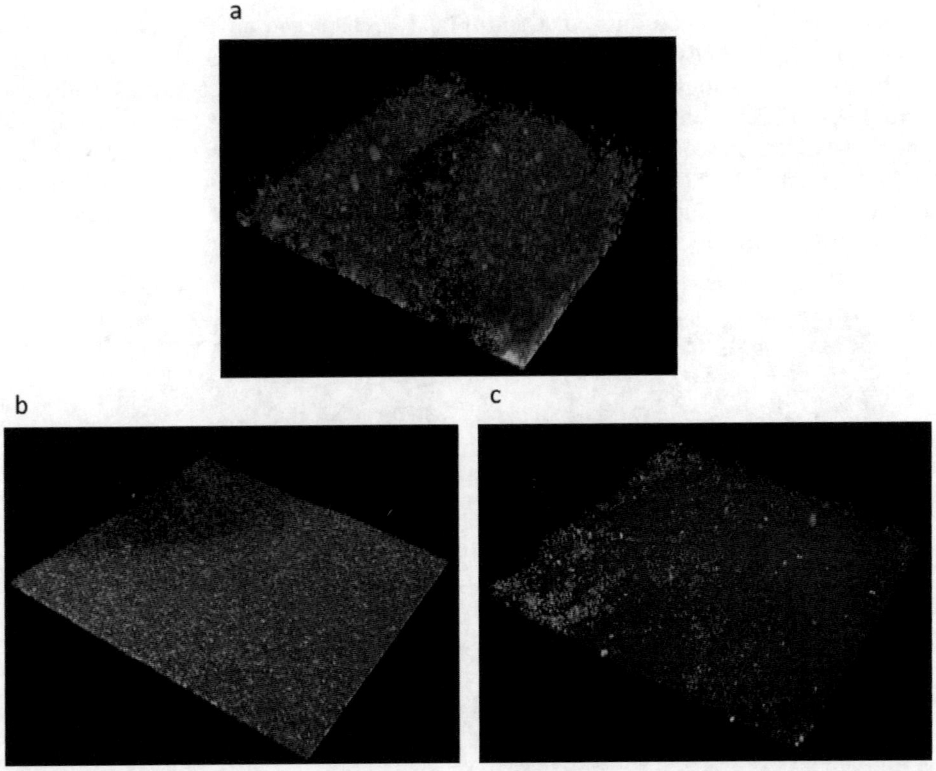

Figure 4. Biofilm growth after 72h in stagnant fluid. Cells were stained with the green fluorescent dye SYTO® 9, while the red fluorescent dye Tetra-methyl-rhodamine (TRITC) conjugated Concanavalin A was used to stain α-mannose residues within the biofilm matrix. Different concentration of non-linked Mag-2 are used: a) Control sample, b) 25μM, c) 50 μM. Image dimension 230x230 μm

This preliminary investigation of the efficiency of Mag-2 shows it is possible to develop a further implementation of the same technology with the goal to realize a nanostructured coating agent on solid surfaces for antibiofilm purpose. The same methodic used to evaluate biofilm morphology under AMP influence in stagnant fluid can be applied for both shaken vessel and in flow setup methods.
Using as control both data represented in Fig.3 and Fig. 4 a comparison between biofilm growth under immobilized-AMPs, non-immobilized AMPs, and absence of AMPs can be made.

3.4. Biofilm morphology under different set-ups

Pseudomonas fluorescens biofilms were firstly formed under the three different conditions described in section 2.3, employing uncoated substrates (Fig. 3), CSLM merged z-stacks images are reported under different flow conditions after 72 h of incubation. Such experiments can be used as control to compare biofilm formation employing the above-mentioned nanostructured coatings.
By visual inspection of the images it can be noticed that different flow conditions induce different morphologies in the biofilm. More specifically, biofilms grown under stagnant conditions (Fig. 3A), presented porous-like structures where isolated cells and colonies are embedded with a uniform EPS matrix. Under shaken fluid conditions (Fig. 3B), biofilms exhibited clumping like structures. In this case, due to the irregularity in the flow, no specific morphological patterns is visible. However, compared to stagnant fluid conditions, biofilms under shaking conditions showed a thinner and heterogenous EPS distribution. Moreover, bacterial cells randomly attached on the substratum (glass case) in clumps structures.

IOP Conf. Series: Materials Science and Engineering 1265 (2022) 012003 doi:10.1088/1757-899X/1265/1/012003

Biofilm grown under well-controlled flow condition (Fig. 3C), displayed different patterns according to the different imposed hydrodynamic conditions. In particular, under low shear rate (Fig 3C, $\dot{\gamma}_w = 0.1\ s^{-1}$), biofilm morphology was comparable to that obtained under stagnant conditions. As shear rate increased, the formation of huge bacterial clusters as well as a reduction of EPS formation was noticeable as the red fluorescent signal considerably reduced (Fig 3C, $\dot{\gamma}_w = 0.4\ s^{-1}$ and $\dot{\gamma}_w = 1\ s^{-1}$). Therefore, it is possible to claim that as shear rate increased, higher biofilm cell density and thickness as well as lower surface coverage were attained as supported by the quantitative investigation of biofilm morphology reported in our previous work [3].

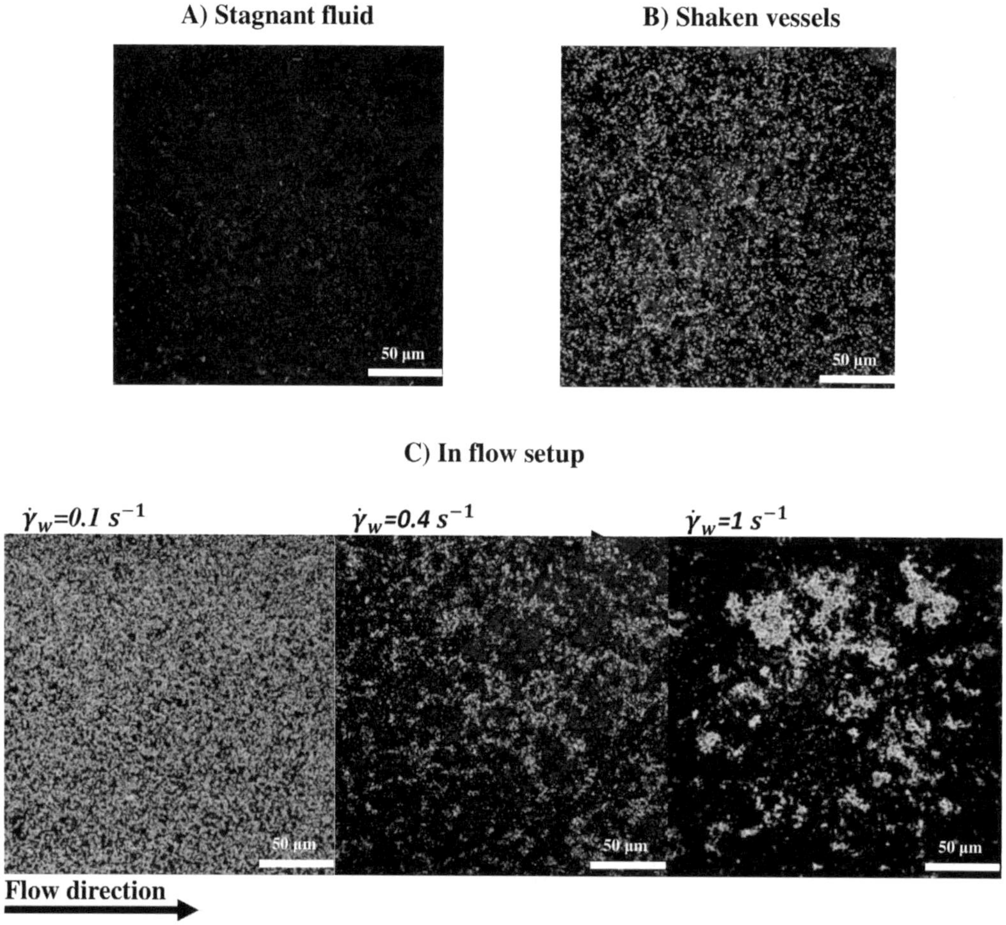

Figure 5. Orthogonal projections of CSLM stack-images, cultivated using the different set-ups proposed. A) stagnant fluid, B) shaken vessel, C) In flow setup ($\dot{\gamma}_w = 0.1\ s^{-1}$, $\dot{\gamma}_w = 0.4\ s^{-1}$, $\dot{\gamma}_w = 1\ s^{-1}$). Bacteria cells are stained with green fluorescent dye, SYTO® 9 and biofilm α-mannose residues are stained with Concanavalin A, in red channel. 40 X magnification, scale bar 50 μm.

4. Conclusions

In this work, we are proposing an innovative experimental approach to investigate AMPs antibiofilm efficiency. The goal of our project is to coat surfaces of selected materials (e.g. glass, metal, polycarbonate) with specific antibiofilm agent. Oxygen plasma technique can be used to bind AMPs on coupon surfaces.

Three different biofilm growth protocols are proposed, corresponding to three different flow conditions (i.e. stagnant fluid, shaken fluid and well-defined laminar flow conditions).

Preliminary tests allowed to identify AMPs, with clear antibiofilm activity, when in solution in the growth medium. Here we reported a preliminary investigation that was used to proof the potential efficiency of the technique proposed.

The final goal of this project is to apply the coating technology on an industrial scale, in order to obtain nano-functionalized surfaces. In particular the project will be focused on the application in the aerospace industry, including application on cabin surfaces and equipment (e.g. water reservoirs) to be used in long range space missions.

Acknowledgements

This study was conducted under the umbrella of the European Space Agency Topical Team: "Biofilms from an interdisciplinary perspective".

Financial support was provided from University of Naples Federico II, Project ROTOR, Finanziamento Ricerca di Ateneo 2021.

References

[1] Di Somma A, Moretta A, Canè C, Cirillo A and Duilio A 2020 l. Antimicrobial and antibiofilm peptides. Biomolecules *Biomolecules* **10** 652.

[2] Flemming H-C and Wingender J 2010 The biofilm matrix *Nature reviews microbiology* **8** 623-33

[3] Recupido F, Toscano G, Tatè R, Petala M, Caserta S, Karapantsios T D and Guido S 2020 The role of flow in bacterial biofilm morphology and wetting properties *Colloids Surf B Biointerfaces* **192** 111047

[4] Frost I, Smith W P J, Mitri S, Millan A S, Davit Y, Osborne J M, Pitt-Francis J M, MacLean R C and Foster K R 2018 Cooperation, competition and antibiotic resistance in bacterial colonies *The ISME journal* **12** 1582-93

[5] Recupido F, Petala M, Caserta S, Kostoglou M, Guido S and Karapantsios T D 2021 Wetting properties of dehydrated biofilms under different growth conditions *Colloids and Surfaces B: Biointerfaces* **210** 112245

[6] Castigliano M, Recupido F, Petala M, Kostoglou M, Caserta S and Karapantsios T D Wetting of Dehydrated Hydrophilic Pseudomonas fluorescens Biofilms under the Action of External Body Forces *Langmuir* **37** 10890-10901

[7] Mattila-Sandholm T and Wirtanen G 1992 Biofilm formation in the industry: a review *Food reviews international* **8** 573-603

[8] Stewart P S and Costerton J W 2001 Antibiotic resistance of bacteria in biofilms *The lancet* **358** 135-8

[9] Høiby N, Bjarnsholt T, Moser C, Bassi G L, Coenye T, Donelli G, Hall-Stoodley L, Holá V, Imbert C and Kirketerp-Møller K 2015 ESCMID guideline for the diagnosis and treatment of biofilm infections 2014 *Clinical microbiology and infection* **21** S1-S25

[10] Cirillo A I, Tomaiuolo G and Guido S 2021 Membrane Fouling Phenomena in Microfluidic Systems: From Technical Challenges to Scientific Opportunities *Micromachines* **12** 820

[11] Persat A, Nadell C D, Kim M K, Ingremeau F, Siryaporn A, Drescher K, Wingreen N S, Bassler B L, Gitai Z and Stone H A 2015 The mechanical world of bacteria *Cell* **161** 988-97

[12] Rajbir S, Debarati P and Rakesh K J 2006 *Trends Microbiol.* **9** 389-97

[13] Feng L, Wu Z and Yu X 2013 Quorum sensing in water and wastewater treatment biofilms *Journal of Environmental Biology* **34** 437-44

[14] Halan B, Buehler K and Schmid A 2012 Biofilms as living catalysts in continuous chemical syntheses *Trends in Biotechnology* **30** 453-65

[15] Wang Z-W and Chen S 2009 Potential of biofilm-based biofuel production *Applied microbiology and biotechnology* **83** 1-18

[16] Di Somma A, Recupido F, Cirillo A, Romano A, Romanelli A, Caserta S, Guido S and Duilio A 2020 Antibiofilm Properties of Temporin-L on Pseudomonas fluorescens in Static and In-Flow Conditions *International journal of molecular sciences* **21** 8526

[17] Di Somma A, Caterino M, Soni V, Agarwal M, di Pasquale P, Zanetti S, Molicotti P, Cannas S, Nandicoori V K and Duilio A 2019 The bifunctional protein GlmU is a key factor in biofilm formation induced by alkylating stress in Mycobacterium smegmatis *Research in microbiology* **170** 171-81

[18] Di Pasquale P, Caterino M, Di Somma A, Squillace M, Rossi E, Landini P, Iebba V, Schippa S, Papa R and Selan L 2016 Exposure of E. coli to DNA-methylating agents impairs biofilm formation and invasion of eukaryotic cells via down regulation of the N-acetylneuraminate lyase NanA *Frontiers in Microbiology* **7** 147

[19] Gordon Y J, Romanowski E G and McDermott A M 2005 A review of antimicrobial peptides and their therapeutic potential as anti-infective drugs *Current eye research* **30** 505-15

[20] Bagheri M, Beyermann M and Dathe M 2009 Immobilization reduces the activity of surface-bound cationic antimicrobial peptides with no influence upon the activity spectrum *Antimicrobial agents and chemotherapy* **53** 1132-41

[21] Godoy-Gallardo M, Mas-Moruno C, Fernández-Calderón M C, Pérez-Giraldo C, Manero J M, Albericio F, Gil F J and Rodríguez D 2014 Covalent immobilization of hLf1-11 peptide on a titanium surface reduces bacterial adhesion and biofilm formation *Acta biomaterialia* **10** 3522-34

[22] Silva R R, Avelino K Y, Ribeiro K L, Franco O L, Oliveira M D and Andrade C A 2016 Chemical immobilization of antimicrobial peptides on biomaterial surfaces *Front. Biosci* **8** 129-42

[23] Bormashenko E, Whyman G, Multanen V, Shulzinger E and Chaniel G 2015 Physical mechanisms of interaction of cold plasma with polymer surfaces *Journal of Colloid and Interface Science* **448** 175-9

[24] Monetta T and Bellucci F 2014 Strong and durable antibacterial effect of titanium treated in Rf oxygen plasma: Preliminary results *Plasma Chemistry and Plasma Processing* **34** 1247-56

Mercapto-Benzamide Inhibitors effects on HIV NCp7 Protein: a parameter-free DFT based structural study

G Cappellini[1,2], R Cardia[1,3], M Valentini[3], E Pieroni[3]
[1]Dipartimento di Fisica, Università di Cagliari, 09042, Cagliari, Italy
[2]ETSF and Dipartimento di Fisica, Università di Cagliari, 09042, Cagliari, Italy
[3]CRS4, Modeling and Simulation Group, Bioscience Department, 09010, Pula, Italy

Abstract. The action of the Mercapto-Benzamide (MB) class of molecules on the HIV Nucleocapsid protein (HIV NCp7), a zinc finger protein, is an issue of relatively recent research interest, relevant to develop a new class of effective and well tolerated HIV antivirals, able to overcome virus escape strategies. MB molecules are easily and cheaply synthesized, and show the ability to unfold the HIV Zinc-finger region, thus avoiding effective viral replication. This effect is not still fully understood, and moreover is highly influenced by the precise composition of MB aromatic ring and chain. Our approach to this biological problem is to adopt a quantum parameter-free (ab-initio: AI) geometrical scheme based on density functional theory (DFT) for the treatment of the electronic degrees of freedom to study with atomistic resolution the action mechanism of MB molecules on NCp7. In particular with respect to the role played by each MB functional group.We report and discuss the outcomes of the here proposed DFT simulations with respect to the different final configurational structures obtained.

1. Introduction

HIV is still a major public health issue with more than 37 million people affected globally [1,2]. The therapeutical schemes being used to attack HIV infection inhibit the essential viral enzymes reverse transcriptase, protease, orintegrase [3].
The principal healing treatment for HIV is the highly active antiretroviral therapy (HAART), in which combinations of drugs simultaneously can target these viral enzymes [4]. Even if infected individuals who adhere to HAART can expect a normal life span, HAART requires regular and lifelong access to costly medication that often impedes many HIV-infected people from receiving the correct treatment. In addition HAART is affected by viral resistance [5] and may also lead patients to cardiovascular and neurological diseases [6,7].
As a consequence of the above points, it is important to continue the study and development of affordable and mutation-resistant small-molecule inhibitors of HIV. Nucleocapsid protein 7(NCp7, (see Figure 1), an 55-residue protein containing two highly conserved zinc-knuckle motifs,[8] results as an attractive target for antiretroviral drugs due to its essential role in viral replication and maturation [9]. NCp7 facilitates essential replication steps, such as strand transfers during reverse tran-scription of viral RNA and genome packaging during virion assembly [10].
In general, NCp7 has proven to be a difficult drug target [11], as there is no distinct three-dimensional structure for the protein other than the zinc-coordinated regions.
Furthermore, traditional medicinal chemistry approaches to target the active sites of enzymes are not applicable to a protein like NCp7. As a result there are currently no NCp7 inhibitors that are used clinically to treat HIV infection.
Some resarch groups are interested in the capacity of mercaptobenzamide(MB)-based molecules to

inhibit HIV-1 through the inactivation of NCp7 [12].
In this work we perform an AI structural minimization, based on DFT for the treatment of the electronic degrees of freedom, in order to study the action of the Mercapto-Benzamide (MB) class of molecules on the HIV Nucleocapsid [12] protein NCp7, a zinc finger protein [13].
More specifically, stimulated by some studies on animal models, we were interested in the impact of the binding of these molecules to the structure of the C-terminal HIV NCp7 "zinc-knuckle" motif [14,15,16]. This motif binds and stabilizes the viral RNA and is thus essential for viral replication and maturation [12,17,18]. Due to its relevance for virus reproduction, the zinc-knuckle is little or nothing muted across HIV strands and undergoes relatively little mutation upon host infection. Therefore, inhibition of NCp7 is a very attractive antiretroviral mechanism, able to escape viral mutation strategies even upon host infection [12,19,20,21]. Unfortunately, NCp7 is a difficult biological target and traditional medicinal chemistry approaches proved to be ineffective [12-15].
On the other hand, MB molecules are easily and cheaply synthesized, well tolerated and, as mentioned, some recent studies found they could be effective inhibitors of the Zinc-finger region, more precisely with a range of IC50 values ranging from 1 to 100 μM, strongly depending on very subtle changes to the aromatic ring and the MB molecular chain [12].
The precise mechanism of action of MB molecules is not fully understood, and the strong dependence on little changes of the atomic structure of the molecules remains definitely unclear. Moreover some MB forms are active in vitro while they lose their effectiveness in vivo and this fact is too yet poorly understood [15]. Fortunately, a 3D crystal structure of the two coordinated zinc-knuckles regions exists [22], that can be used to build the Zinc-finger motif model.
We adopted a modeling hypothesis, explained in the Numerical Methods section, that allowed us to chop the problem in simpler steps, in such a way to mimic complex biological mechanisms and such that each step can be simulated with the appropriate method.
The expected results will be fundamentals to design the molecule pharmacological activity.According to the existing literature, our hypothesis is that in the body biological environment, the MB2 molecules are partly acetylated to the form AcMB2, where the aromatic ring sulfhydryl group (SH) is replaced by the acetyl group ($SOCH_3$) (Figure 2). The acetylation of MB2 (a particular important form of MB molecules[12]) guarantees survival and migration of AcMB2 inside the cell, where it will eventually meet the NCp7 protein. Then, when the AcMB2 molecule interacts with the NCp7 protein, the acetyl group is transferred to NCp7 CYS413 (Figure 2), exposing the AcMB2 S ion, that is now free to interact with the motif coordinating Zn and eventually disrupts Zn coordination of the remaining CYS416, CYS426 and HIS421, leading to protein misfolding or even unfolding. Therefore we believe the active form of the MB2 molecule is precisely the AcMB2 molecule without the acetyl group and with the S ion exposed, indicated in the following as MB2act.

2. Modeling Protocol

According to our work hypothesis, we developed the following simplified modeling protocol:

- Protein model: we selected model number 1 from PDB NMR structure 2L44 [22] of the C-terminal zinc knuckle of the HIV NCp7. The structure is composed of 18 aminoacids representing the zinc knuckle motif Cys3-X2-Cys6-X4-His11-X4-Cys16 corresponding to NCP7 protein [Uniprot Q9YP46] aminoacids 411-429.
- AI DFT-based quantum simulation to optimize the MB2act molecular structure, and its electronic characterization (orbitals, ground state, ...).
- AI DFT-based quantum simulation of MB2act interacting with the Zn-finger motif in vacuum and with implicit water. The MB2act sulphur atom has been placed close to the zinc coordination atom.
- Analysis of the distance variations between the cysteine sidechain sulphur atoms, histidine nitrogen atom and the central coordinating zinc atom during the DFT runs is performed. To evidence structural changes, particularly motif unfolding, we also analysed the variations of distances between the NCp7 backbone Calphas relative to the histidine and cysteines that define the motif.

3. AI DFT Structural Simulations

We perform an AI DFT based [23,24] structural relaxation using the Berny algorithm [25-b] (in order to find the true local minimum on the potential energy surface of the system) as implemented in the Gaussian16 simulation package [25]. Geometry optimizations have been obtained using the Becke three-parameter Lee–Yang–Parr (B3LYP) hybrid electronic exchange–correlation (XC) functional [26,27], in combination with the 6–31 G* basis-set, a valence double-ζ set augmented with d polarization functions. The B3LYP electronic exchange-correlation (XC) functional has stable behavior, with some well-documented limitations [28]. The structural optimization of the system was performed without imposing symmetry constraints. The procedure here followed has been successfully tested for different systems, ranging from organic to inorganic clusters of different dimension and grade of complexity [29-33]. The calculation of the exact ground state is also of fundamental importance for systems of biological interest as the first step that should be accomplished for the calculation of electronic excitations and optical properties [34,35]. Notice that, being mainly interested to the protein structure evolution, we perform simulations with a total number of atoms quite larger than that used for a standard structural optimization. Moreover we performed all the main simulations using the SMD implicit solvent model [25-c] in order to take in account at least the main effects of the typical water environment in the living cells. In conclusion in this project the motivation to use the AI DFT scheme described above instead of a Molecular Dynamics (MD) one has been mainly related to the dimension of the studied protein-related problem which could be tackled safely within a parameter-free method, to the possibility to follow the creation and breaking of chemical bonds during the dynamics and to the possibility to find the real conformation of minimum energy of the studied system.

4. Results and Discussion

In the Figure 1 is reported within two different perspectives the Nucleocapsid protein 7(NCp7) which is the target of the present study. Reported in the right panel are also the main groups forming the protein, namely CYS426, HYA421,CYS416,CYS413.

In Fig.2 on the right panel the Mercaptobenzamide inhibitor is reported and in the left panel the final structure of unfolded NCp7 + MB2 inhibitor (considered in the implicit water simulation) obtained by the quantum relaxation sheme here used. In the left panel is also highlighted the final distance of Zn-CYS 426 after the structural relaxation.

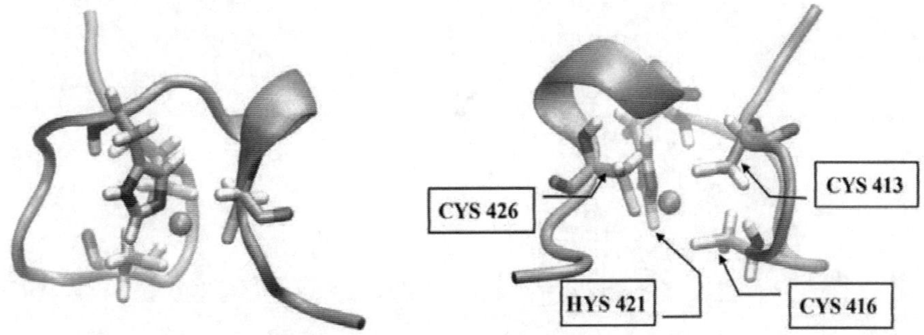

Fig. 1 Left panel Ncp7 secondary and tertiary structure. The 4 residues highlighted on the right panel are those forming the C-terminal Zinc-knuckle. Data taken from PDB file [22].

To make a more complete investigation we have also performed relaxation in case of a non active MB2 inhibitor. In Fig.(3) the outcome of that simulation are reported in the right panel. In this case the effect of the depontiated MB2 inhibitor is clearly visible in relation to the reduced opening of the zinc-

knuckle.
To have a quantitative insight in the effects of the MB2 inhibitor we report in Tab.(1) the distances between the Zn atom the groups of the NCp7 protein in the different simulation schemes. In the first column the experimental data after Ref. (22), in the second column the results of the simulation performed aftyer the optimization in vacuum. In paretheses the percentage of deviations with respect to the experiment.
The distance between the Zn atom and the CYS 426 group shows a consistent enlargement (more than 30%). In the third is present also the MB2 inhibitor. In this particular case the above distance is enlarged by more a factor 5, demonstrating the active action of the MB2 inhibitor and the unfolding of the Zn-knuckle. The consequence of the presence of real water molecules in the simulatioin is clearly highlighted in the values of the distances reported in the fourth column of this table. The distance between Zn atom and the CYS 426 group shows an enlargement with respect to the experiment. Interestingly, this enlargement is strongly hampered in case the simulation is started from the use of an MB2 inactivated inhibitor. The Zn-Cys 426 distance retourns to a value comparable with

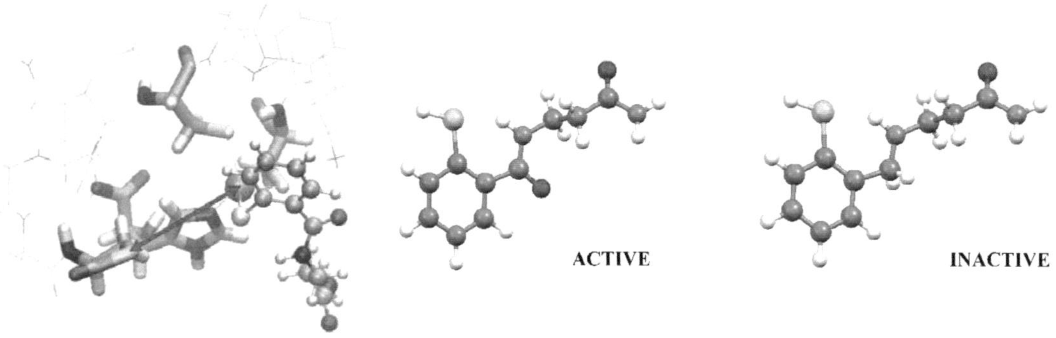

that obtained for the free NCp7 protein in vacuum (see Fig.(3)).

Fig. 2 On the left the final structure of unfolded NCp7 + MB2 inhibitor (implicit water simulation). Highlighted the final distance of Zn-CYS 426 after the structural relaxation. In the center and right panels the Mercaptobenzamide inhibitor (Active form in center, Inactive form on the right).

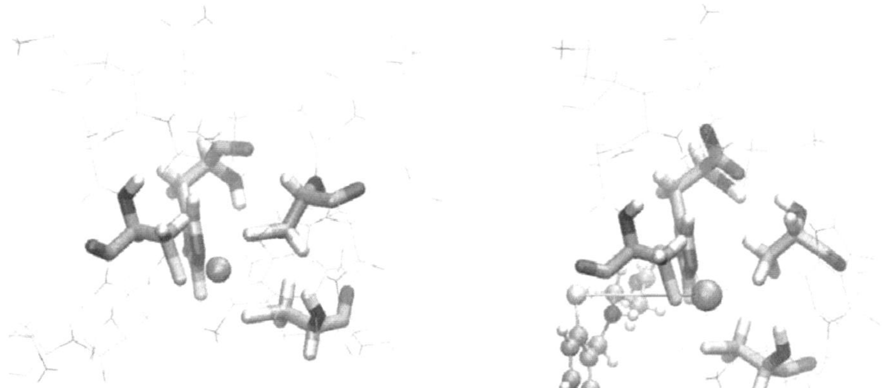

Fig. 3 Comparison between the experimental geometry of NCp7 (left panel) and final simulation frame in the case of non active MB2 inhibitor form (right panel, implicit water simulation).

Table I.The distance between the Zn atom the groups of the NCp7 protein in the different simulation schemes. In the first column the experimental data after reference Ref. (22), in the second column the results of the simulation performed after the optimization performed in vacuum. In the third is present also the MB2 inhibitor. In the fourth the MB2 inhibitor with implicit water (H_2O). In the fifth the MB2 inhibitor modified in the inactive form with implicit water (H_2O).

	NCp7 [Å] (Exp.)	NCp7 [Å] (Opt.Vacuum)	NCp7+MB2 [Å](Vacuum)	NCp7+MB2 [Å](H_2O)	NCp7 + MB2[Å] (Inactive MB2 - H_2O)
$d_{Zn\_Cys\ 413}$	*2.30*	*2.34* (+ 1.2 %)	*2.42* (+ 5.2 %)	*2.38* (+ 3.5 %)	*2.35* (+ 1.2 %)
$d_{Zn\_Cys\ 416}$	*2.27*	*2.25* (<1 %)	*2.44* (+ 7.5%)	*2.38* (+ 4 %)	*2.40* (+ 6 %)
$d_{Zn\_Cys\ 426}$	*2.37*	*3.17* (+ 34 %)	*13.03* (~ x 5.5)	*8.45* (~ x 3.6)	*3.24* (+ 37 %)
$d_{Zn\_Hys\ 421}$	*2.05*	*1.88* (+8.3 %)	*2.03* (+ 1 %)	*2.03* (+1 %)	*2.04* (<+1 %)

5. Conclusions

The effect of MB molecules upon HIV Zinc-finger region has been analysed within an "ab-initio" DFT based full quantum framework. Our simulations demonstrate that the main effects of the MB2 inhibitor is to unfold the Zn-knuckle of the on NCp7 protein, even in the presence of water. This result can be hampered by small structural changes in the MB2 inhibitor.
We believe the results presented here and their future development will be useful for rational drug design, including discovering interesting targets and understanding molecular action mechanisms.
Due to recent emerging results we are planning to continue and deepen this line of research by explicitly including the acetyl group and by MD simulation techniques to explicitly solvate the system and elucidate distinct protein equilibrium states, with potentially distinct roles on antiviral action mechanism.

6. Acknowledgements

The authors acknowledge the use of the ISCRA-CINECA facility within the ISCRA-C 2021 project HP10CDZJIB. G.C. acknowledges collaboration with "Progetto biennale d'Ateneo UniCa finanziato dalla Fondazione di Sardegna annualità 2020 : Molecular simulations and machine learning in bio- and medical physics: from antimicrobials to radiopharmaceuticals" and partial financial support from IDEA-AISBL-Bruxelles.

7. Bibliography

[1] AIDS by the numbers 2015,UNAIDS, November24, 2015:www.unaid-s.org/en/resources/documents/2015/AIDS_by_the_numbers_2015.
[2] Eriksen J et al. 2016 *Infect. Dis.* 1–34
[3] a) de Bethune M P 2010 *Antiviral Res.* **85** 75–90
b) Prokofjev M M et al. 2016 *Acta Naturae* **8** 23–32
c) Palomba M et al. 2016 *Org. Biomol. Chem.* **14**, 2015 –2024
d) Corona A et al. 2016 *ChemMedChem* **11** 1709 –1720
[4] Moore R D, Chaisson R E 1999 *AIDS* **13** 1933–1942
[5] Cortez K J, Maldarelli F, 2011 *Viruses* **3**, 347 –378
[6] Broder S 2010 *Antiviral Res.* **85**, 1–18
[7] Wensing A M et al. 2015 *Top. Antiviral Med.* **23**,132–141
[8] a) Berg J M 1986 *Science* **232** 485–487
b) Cornille F et al. 1990 *Int. J. Pept. Protein Res.* **36** 551–558
c) Morellet N et al. 1992 *EMBO J.* **11** 3059 –3065

[9] a) Rice G R et al. 1995 *Science* **270** 1194
b) Domagala J M et al. 1997 *Bioorg. Med. Chem.* **5** 569
c) Elslager E F et al. 1995 Oct 31 (Warner- Lambert Co.), US Pat. No. US5463122 A
d) Bolton G L et al. 1997 April 15 (Warner-Lambert Co.), US Pat. No. US5620997 A
e) Goldschmidt V et al. 2010 *HIV Ther.* **4** 179–198
[10] Muriaux D, Darlix J L 2010 *RNA Biol.* **7** 744 –753
[11] Spearman P 2016 *Curr. Top. Med. Chem.* **16** 1154–1166
[12] Saha M et al. 2017 *ChemMedChem* **12** 714 – 721
[13] Ren W et al. 2018 *PLoS ONE* 13 (5) e0196662
[14] Berg J M 1986 *Science* **232** 485-487
[15] Cornille F et al. 1990 *Int. J. Pept. Protein Res.* **36** 551-558
[16] Morellet N et al. 1992 *EMBO J.* **11** 3059-65
[17] Rice G R et al. 1995 *Science* **270** 1194
[18] Domagala J M et al. 1997 *Bioorg. Med. Chem.* **5** 569
[19] Iraci N et al. 2018 *Drug Discovery Today* **23-3** 687-695
[20] Sancineto L et al. 2018 *Drug Discovery Today*, **23-2,** 260-271
[21] Shvadchak V et al. 2018 *Biochemistry* **57-30** 4562-4573
[22] Quintal S et al. 2012 *Biochemistry* **51 (8)** 1752-1761
[23] Kohn W 1999 *Rev. Mod. Phys.* **71** 1253–1266
[24] Jones R.O., Gunnarsson O 1989 *Rev. Mod. Phys.* **61** 689–74
[25] a) Gaussian 16, Revision C.01, Frisch M J et al. 2016 Gaussian, Inc., Wallingford CT
b) Schlegel H B 1982 *J. Comp. Chem.* **3** 214-18
c) Marenich A V et al. 2009 *J. Phys. Chem. B* **113** 6378-6396
[26] Becke A D 1993 *J. Chem. Phys.* **98** 5648–5652
[27] Lee C et al. 1998 *Phys. Rev.* B **37** 785–789
[28] Grimme S, Parac M 2003 *ChemPhysChem* **4** 292–295
[29] Malloci G et al. 2004 *Phys. Rev.* B *Condens. Matter Mater. Phys.* **70** 1–6
[30] a) Cardia R et al. 2014 *J. Phys. Chem.* A **118** 5170–5177
b) Cardia R et al. 2016 *Phys. Rev.* B **93** 235132
c) Dardenne N et al. 2017 *J. Phys. Chem* C **121 (44)** 24480–24488
d) Antidormi A et al. 2018 *J. Phys. Chem* C **122 (49)** 28405–28415
[31] Cardia R et al. 2016 *Opt. Photonics J.* **6** 41–47
[32] a) Mocci P et al. 2018 *Journal of Physics: Conference Series* **956 (1)** 012020
b) Mocci P et al. 2018 *New Journal of Physics* **20 (11)** 113008
[33] Cappellini G et al. 2020 *ACS Omega* **5 (22)** 13268-13277
[34] Kumar A et al. 2018 Cellulose **25 (4)** 2191-2203
[35] Kumar A et al. 2019 Cellulose **26 (3)** 1489-1501

IOP Conf. Series: Materials Science and Engineering 1265 (2022) 012005 doi:10.1088/1757-899X/1265/1/012005

Ab initio circular dichroism with the Yambo code: applications to dipeptides

Elena Molteni[1,2,3,*], Giancarlo Cappellini[3,4] and Davide Sangalli[1,2,3]

[1] Istituto di Struttura della Materia - CNR (ISM-CNR), Division of Ultrafast Processes in Materials (FLASHit), Area della Ricerca di Roma 1, Via Salaria km 29.300, CP 10, Monterotondo Scalo, Roma, Italy

[2] Department of Physics, Università degli Studi di Milano, via Celoria 16, I-20133 Milano, Italy

[3] European Theoretical Spectroscopy Facility (ETSF, www.etsf.eu)

[4] Department of Physics, Università degli Studi di Cagliari, and CNR-IOM SLACS Cagliari, Cittadella Universitaria di Monserrato, 09042 Monserrato, CA, Italy

E-mail: * elena.molteni@mlib.ism.cnr.it

Abstract. Circular dichroism (CD) spectroscopy is a useful technique for characterizing chiral molecules. It is more sensitive than total absorption to molecule conformation, and it is routinely used to identify enantiomers. We present here a first principles implementation of CD with application to three cyclo-dipeptides. Our CD approach for molecules has been integrated in the 5.0 release of the Yambo code, distributed under GPL.

1. Introduction

Circular dichroism (CD) spectroscopy, due to the different absorption of left *vs* right circularly polarized light by chiral systems, is a useful technique for characterizing chiral molecules, it can be used to identify different enantiomers, and it is also able to yield information on molecule conformation (*e.g.* [1; 2] and refs. therein). Chirality plays an important role in molecular recognition; therefore it is of interest in several fields, from drug discovery to catalysis to biomolecular function. Computational investigations of the CD spectra of molecules can help in several ways: they can allow to predict which spectral regions are more sensitive to the absolute configuration (enantiomer) of a given molecule, and which ones to its conformation; if a chiral drug molecule yields a clear CD "fingerprint" (i.e. well recognizable features in the CD spectrum), this may be used *e.g.* for assessing its accumulation in cells.

We report here on our implementation of CD calculations within the Density Functional Theory (DFT) framework in the Yambo code [3], and on its application to computing (absorption and) CD spectra of three cyclo-dipeptides, cyclo(Glycine-Phenylalanine), cyclo(Tryptophan-Tyrosine) and cyclo(Tryptophan-Tryptophan), of which some of us had previously characterized the electronic occupied states[4].

Cyclo-dipeptides (CDPs) or 2,5-diketopiperazines (DKP) are interesting both thanks to their biological and pharmacological activities (such as antibacterial, antiviral, antitumoral, antioxidant)[5] and as possible building blocks for nanodevices[6; 7] due to their multiple hydrogen bonding sites, which can potentially have a role in self-assembly[6; 8]. As chiral molecules, CDPs can catalyze enantioselective reactions[9]; they have been detected e.g. in meteorites[10] and can considered as precursors of longer peptides[10; 11]: they may therefore have had a role also in the "homochirality" of life, *i.e.* the prevalence of the L enantiomer of amino acids in proteins[10]. Moreover, the role of molecule chirality in self-organization of cyclo-dipeptides has also attracted interest[7].

2. Methods

Methyloxirane is first used as a reference molecule to validate the approach, and then we proceed to compute optical properties of the three cyclo-dipeptides. Optical absorption and circular dichroism (CD) spectra are computed within the Independent Particle (IP) approximation with the Yambo code [3]. Yambo is a plane wave code interfaced with QuantumESPRESSO (QE) [12; 13] that allows calculation of the optical response starting from the previously generated Kohn-Sham (KS) wave functions and energies in a plane-wave basis set. In the QE calculations the molecule (either methyloxirane or a cyclo-dipeptide) was put in a face-centered cubic (FCC) cell with lattice parameter $a = 56.087$ a.u. (*i.e.* ≈ 29.68 Å). An FCC cell yields the same distance between molecules in replicated cells with a smaller volume, with respect to a simple cubic cell, and, therefore, a lower number of G-vectors, which makes the calculation less computationally expensive. Total energies have been calculated using norm-conserving Troullier-Martins atomic pseudopotentials[14]. The LDA approximation for the exchange-correlation potential is used for methyloxirane, to allow comparisons with previous computational works on optical spectra of this molecule[2; 15]. The hybrid B3LYP [16; 17] functional, which is widely used for molecules and has been shown also recently by some of us[4] to yield a good description (up to a rigid energy shift) of the valence electronic structure of the three cyclo-dipeptides c-GlyPhe, c-TrpTyr and c-TrpTrp, was instead chosen for these latter molecules. The D3 pairwise dispersion correction for van der Waals (vdW) interactions [18] was also included for relaxing the geometry of the three cyclo-dipeptides, and the Makov-Payne correction to the total energy was used to compute the vacuum level, and to properly align electronic energy levels [19]. In Yambo calculations we used 200 bands for methyloxirane, and 500 bands for the three investigated cyclo-dipeptides. In all cases the same plane wave cutoff for wavefunctions of 45 Ha gives a good convergence of the energy levels, total energies and atomic forces in the QE runs.

In our approach, both absorption and CD are constructed starting from the matrix elements of the position operator which, for isolated molecules, we compute in real space

$$\mathbf{r}_{nm} = \langle \psi_n | \mathbf{r} | \psi_m \rangle \tag{1}$$

where $\mathbf{r}$ is a vector of component r_j, with $j = x, y, z$. We define $\omega_{nm} = (\epsilon_n - \epsilon_m)/\hbar$, the excitation frequency for an electronic transition between the states ψ_n and ψ_m. From the latter we define the velocity dipoles $\mathbf{v}_{nm} = \mathbf{r}_{nm}\omega_{nm}$, and the magnetic dipoles

$$\mathbf{m}_{mn} = \sum_{l=l_{min}}^{l_{max}} \mathbf{r}_{ml} \times \mathbf{v}_{ln}. \tag{2}$$

The expression for the magnetic dipoles results from the use of the identity operator $\sum_l |l\rangle\langle l|$. Hence the n and m indexes belong to a given transition (from an occupied to an unoccupied orbital in the resonant case) while the l index should run over all orbitals, i.e. $l_{min} = 1$ and $l_{max} = \infty$. In practice one should verify the convergence of calculated CD spectra, in a given energy range, both on the transitions included, and on the l index.

For Independent Particle (IP) absorption spectra we calculated the polarizability α, in terms of the sum of direct transitions between Kohn-Sham eigenstates within the Fermi Golden Rule

$$\alpha_{ij}(\omega) = -4\pi \sum_{nm} \left(\frac{r^i_{nm} r^j_{mn}}{\omega - \omega_{nm} - i\gamma} + \frac{r^j_{nm} r^i_{mn}}{\omega + \omega_{nm} + i\gamma} \right). \tag{3}$$

The IP circular dichroism signal, within linear response, is instead proportional to the G-tensor[2; 20; 21]:

$$G_{ij}(\omega) = \frac{q_e^2}{2m\hbar} \sum_{nm} \left(\frac{r^i_{nm} m^j_{mn}}{\omega_{nm} - \omega - i\gamma} + \frac{m^j_{mn} r^i_{nm}}{\omega_{nm} + \omega + i\gamma} \right) \quad (4)$$

For randomly oriented chiral molecules absorption and CD are expressed as the trace of the $\alpha(\omega)$ and $G(\omega)$ tensors, respectively.

3. Results

3.1. Numerical tests on R-methyloxirane and convergence in c-GlyPhe

Methyloxirane has often been used in the literature as a benchmark molecule for CD calculations against experimental data, due to its rigidity. For flexible molecules instead, one has to take into account the fact that the different possible conformers will in general yield different CD spectra, which makes a comparison to experimental spectra not trivial. In Figure 1 we report the geometry (left panel) of R-methyloxirane (*i.e.* the "R" enantiomer of methyloxirane) and its calculated absorption (panel a) and CD (panel b) spectra, compared to the vacuum UV experimentally measured absorption and CD spectra of the same molecule[22].

Our absorption and CD spectra of R-methyoxirane, calculated at IP level, reproduce well the first two features of the corresponding experimental spectra[22] provided a rigid shift of 1.4 eV is applied to calculated ones. The agreement is of the same quality of that reported in previous computational works[2; 15]. The discrepancies between theory and experiment in the high-energy part of the spectra have also been reported in the literature and they may be due to the IP approximation[2].

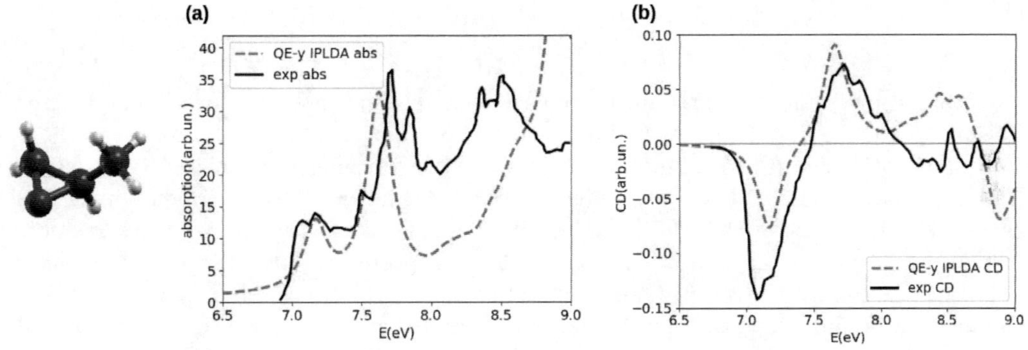

Figure 1. Left panel: Geometry of R-methyloxirane. Panels (a) and (b): IP (dashed red line) absorption (panel a) and CD (panel b) calculated spectra of R-methyloxirane, obtained within DFT LDA, compared to its experimental spectra (solid black line). Calculated spectra have been shifted by +1.4 eV, and a broadening of 0.1 eV has been used.

One of the important aspects of CD implementations, is that it involves the definition of the orbital magnetic dipoles, which are ill defined in periodic boundary conditions. In our approach to the orbital magnetic dipoles, this problem appears in the presence of the terms $\mathbf{r}_{nn}$ in eq. 2. In isolated systems this is not an issue, since $\mathbf{r}_{nn}$ can be directly evaluated in real space. On the other hand in extended systems the position operator is ill defined and only components between non degenerate states can be computed, via the evaluation in reciprocal space of $\mathbf{v}_{nm}$, and later using $\mathbf{r}_{nm} = \mathbf{v}_{nm}/\omega_{nm}$. Instead $\mathbf{r}_{nm} = 0$ must be imposed if $\omega_{nm} < E_{thresh}$. Here we have verified, by comparing CD spectra of R-methyloxirane obtained by computing dipoles either in real space or in reciprocal space (data not shown) that the $\mathbf{r}_{nn}$ dipoles have a negligible effect on computed CD spectra. This suggests that our approach, here implemented for molecules computing dipoles in real space, may be successfully extended to the case of solids, where the G space approach is generally used.

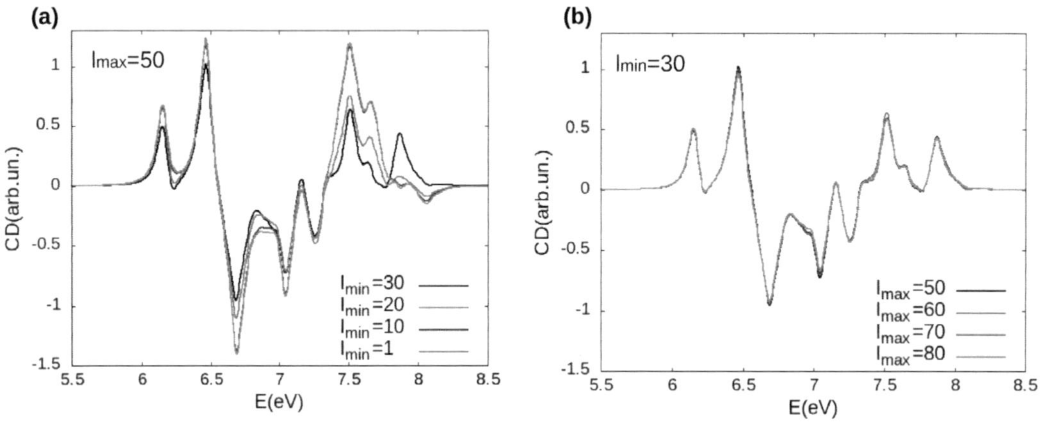

Figure 2. Independent Particle CD spectra of the c-GlyPhe dipeptide, obtained with different values for the the Yambo `DipBands` keyword, used to converge the identity resolution entering the definition of the magnetic dipoles **m**. Convergence is verified varying the value of l_{min} (left panel), and l_{max} (right panel) independently. The molecule has 39 occupied states.

After validating on R-methyloxirane our computational scheme for (absorption and) circular dichroism spectra, we first consider the lowest energy conformer (see also discussion in the next section) of cyclo(Glycine-Phenylalanine) to verify the convergence of calculated CD spectra on the number of states used to resolve the identity in the definition of the magnetic dipole matrix elements (see discussion in the Methods section, l index in eq. 2). Convergence results are presented in Fig. 2. The CD spectrum of c-GlyPhe is reported at fixed number of transitions, using in Eq. 4 10 occupied states (index m ranging from 30 to 39) and 11 empty states (index n ranging from 40 to 50), while varying the number of states included in the sum over the l index. In the left panel we consider the convergence with respect to the value of l_{min} (index in the occupied states), while in the right panel we consider the convergence with respect to the value of l_{max}. The two parameters, i.e. number of transitions and number of states included in the resolution of the identity, are controlled independently in the Yambo input file *via* the use of the two variables `BSEbands` and `DipBands` respectively. Magnetic dipoles, and hence CD spectra, are weakly sensitive to the range of empty states used in the expression of $\mathbf{m}_{mn}$, while the convergence on occupied states is less trivial, requiring the inclusion of occupied states down to state 10 (the molecule has 39 occupied states) for a reasonable spectrum. Here the CD spectra are computed via the real space procedure for the dipoles, since the reciprocal space dipoles would require a direct evaluation of the commutator with the non local part of the B3LYP exchange and correlation potential.

3.2. CD spectra of Cyclo-dipeptides

We now report results on three cyclo-dipeptides with aromatic sidechains, namely cyclo(Glycine-Phenylalanine), cyclo(Tryptophan-Tyrosine) and cyclo(Tryptophan-Tryptophan). At a

IOP Conf. Series: Materials Science and Engineering 1265 (2022) 012005 doi:10.1088/1757-899X/1265/1/012005

difference with the above-discussed methyloxirane, the three chosen cyclo-dipeptides (c-GlyPhe, c-TrpTyr, c-TrpTrp) display some flexibility; therefore, also in view of possible comparisons to experimentally measured absorption and/or CD spectra, one should look for the most stable conformers, which are expected to be the most abundant ones in experiments, apart from possible effects of the experimental conditions (solvent vs. gas phase, temperature, etc.). For each of the three investigated cyclo-dipeptides, therefore, we have considered the lowest energy gas phase conformers as obtained by some of us in a previous work through a tight-binding conformational search, followed by geometry optimization within B3LYP DFT[4]. These conformers are shown in the left panels of Figs. 3, 4 and 5 for c-GlyPhe, c-TrpTyr and c-TrpTrp respectively, nearby the IP absorption and CD spectra.

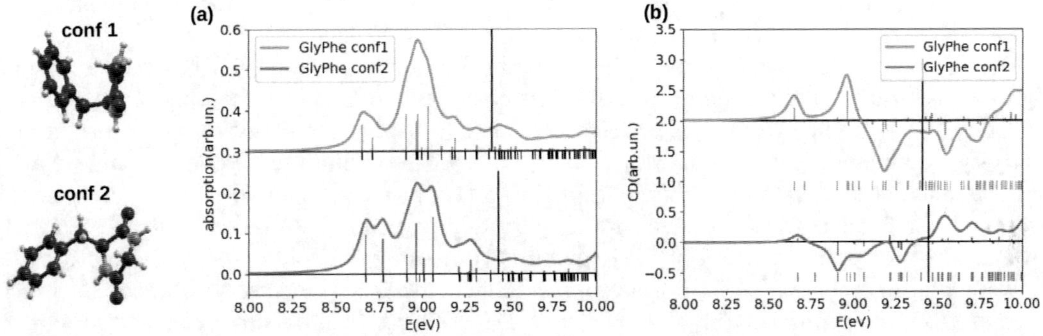

Figure 3. Left panel: geometry of the chosen low-energy conformers of c-GlyPhe. Panel (a): Independent Particle (IP) B3LYP absorption spectra of c-GlyPhe conformer 1 (magenta) and conformer 2 (gray). Panel (b): IP B3LYP circular dichroism (CD) spectra of the same two conformers (same color codes). Vertical black lines indicate E(vacuum level) - E(HOMO) for the two conformations. A shift of +2.5 eV has been applied to absorption and CD spectra, and a broadening of 0.05 eV was used.

In the IP approximation excited states correlations, which would lower the optical gap with respect to the electronic gap $E_{LUMO} - E_{HOMO}$, are neglected. On the other hand, the electronic gap $E_{LUMO} - E_{HOMO}$ is underestimated by standard DFT functionals, such as LDA or B3LYP. The two mentioned effects are of opposite sign, therefore the IP-calculated energy position of the absorption onset can be either over- or underestimated with respect to the experimentally observed one, depending on which of the two effects is larger for the specific molecule under study. In the analysis of the spectra of R-methyloxirane, the sum of the two errors results in an underestimation of the optical gap. A simple rigid shift of +1.4 eV to the calculated IP spectrum was enough to give a very good description of CD. For cyclo-dipeptides instead we fix the HOMO - LUMO gap with a shift of +2.5 eV, based on a previous study [4], where measured photoemission spectra (PES) are carefully compared with *ab initio* simulations. We are thus left with an overall overestimation of the position of the optical gap. A direct comparison of IP results to experimentally measured spectra is not the main goal of the present work. This is why for all the three investigated cyclo-dipeptides we did not look for the additional shift needed to match the position of their optical spectra. In the case of the widely studied Trp amino acid[23; 24], and also of the c-TrpTrp dipeptide[25], experimentally measured absorption spectra display a strong absorption peak in the 4 to 5 eV energy region, whose maximum lies at lower energy with respect to our IP calculated spectra, even before applying the above-mentioned +2.5 eV shift to them. An investigation of absorption and CD spectra of these molecules beyond

IP level may be the subject of further works.

On the other hand the calculated IP spectra reported here yield information on the contribution of individual state to state transitions to the intensity of absorption and CD peaks *via* the dipole matrix element between the two involved states, whose intensity can be strongly affected by the localization of these states (including the possible absence of specific peaks corresponding to dipole-forbidden transitions). This information is interesting when comparing spectra of different molecular conformers, and it adds to our knowledge of the electronic properties of these cyclo-dipeptides, with respect to the simple picture obtained by DFT electronic densities of states[4] which only depend on the energy distribution of electronic states. Having aligned the transition energies with the position of the energy level measured in photo-emission, helps in mapping a peak to the associated occupied and empty states.

In Fig. 3 panels (a) and (b) we compare Independent Particle (IP) B3LYP optical absorption and electronic circular dichroism (CD) spectra of conformers 1 and 2 (left panel of the same figure) of the c-GlyPhe peptide, obtained with the Yambo code from QE KS wavefunctions. In spite of the quite similar energy distribution of electronic levels [4] for the two considered conformers of c-GlyPhe, absorption spectra (panel (a)) display some degree of conformational sensitivity: the main features lie in both cases around 8.7 eV and around 9 eV, but their relative intensities and detailed shapes are different for the two conformers: in conformer 2, at a difference with conformer 1, each of these two features is splitted into two peaks. Conformational sensitivity is even more pronounced in CD spectra (panel (b)): here corresponding spectral features for the two conformers have in several cases opposite signs, thus yielding very different dichroism spectra. This strong conformational dependence of CD spectra has already been reported in the literature for single amino acids (see [2] and refs therein).

Table 1. Energies (shifted by +2.5 eV), identity of the involved valence to conduction transition, and contribution to absorption and CD for low energy transitions in conformers 1 and 2 of c-GlyPhe: asterisks indicate transitions contributing to absorption (column 3) or CD (column 4) peaks.

c-GlyPhe conf1 E(eV)	transition	abs	CD
8.65	H - L	*	*
8.71	H - (L+1)	*	
8.91	(H-1) - L	*	*
8.97	(H-1) - (L+1)	*	
8.98	(H-2) - L	*	
9.00	H - (L+2)		
9.04	(H-2) - (L+1)	*	
c-GlyPhe conf2 E(eV)	**transition**	**abs**	**CD**
8.67	H - L	*	*
8.77	H - (L+1)	*	
8.91	(H-1) - L	*	*
8.97	(H-2) - L	*	
8.99	H - (L+2)		
9.01	(H-1) - (L+1)	*	*
9.07	(H-2) - (L+1)	*	*

In c-GlyPhe conformation 1 (magenta curve in panel (a) of Fig. 3) six out of the first (in energy order) seven transitions between Kohn-Sham states (short vertical black ticks below the absorption spectrum), namely all of them except the sixth one, *i.e.* the HOMO - (LUMO+2) one, give non-negligible contributions (vertical magenta ticks) to the Independent Particle absorption spectrum (see also Table 1, third column). In particular, the first and second transitions contribute to the absorption peak at $\approx$ 8.7 eV, while the 3rd, 4th, 5th and 7th transitions contribute to the absorption peak at $\approx$ 9 eV. The fact that most low-energy transitions give non-negligible contributions to the IP absorption spectrum is in agreement with the observation that most of the highest occupied and lowest unoccupied electronic states in this system[4] are not localized in a specific part of the molecule (in that case low intensity contributions would be expected to occur for transitions between pairs of states localized on separated and relatively "far" parts of the molecule). Also in c-GlyPhe conformation 2 (grey curve in panel (a) of Fig. 3) six out of the first seven transitions (short vertical black ticks) are bright (vertical grey ticks in the Figure, third column of Table 1). Again only the HOMO - (LUMO+2) transition is dark. For both conformers all the mentioned bright low energy transitions have either the LUMO or the LUMO+1 as "final" conduction state. Only a subset of the optically bright transitions give non-negligible contributions to the corresponding CD spectra (panel (b) of Fig. 3, fourth column of Table 1), *i.e.* only two (respectively four) transitions in c-GlyPhe conformation 1 (conformation 2, respectively).

In Fig. 4, panels (a) and (b), we report the Independent Particle B3LYP absorption (a) and circular dichroism (b) spectra of the three lowest energy conformers (left panel of the same figure) of c-TrpTyr. Although these three absorption spectra share several common features, they can be distinguished from each other. For c-TrpTyr conformation 1 (magenta lines in Fig. 4) the only relevant contribution to the first IP absorption peak stems from the HOMO-LUMO transition (7.46 eV, see also third column of Table 2); the main contribution to the second absorption peak comes from the (HOMO-1) - LUMO transition, with a less intense contribution from the (HOMO-2) - LUMO one; the (HOMO-2) - (LUMO+1) transition, of comparable intensity as the (HOMO-2) - LUMO one, gives rise to the shoulder of the second absorption peak.

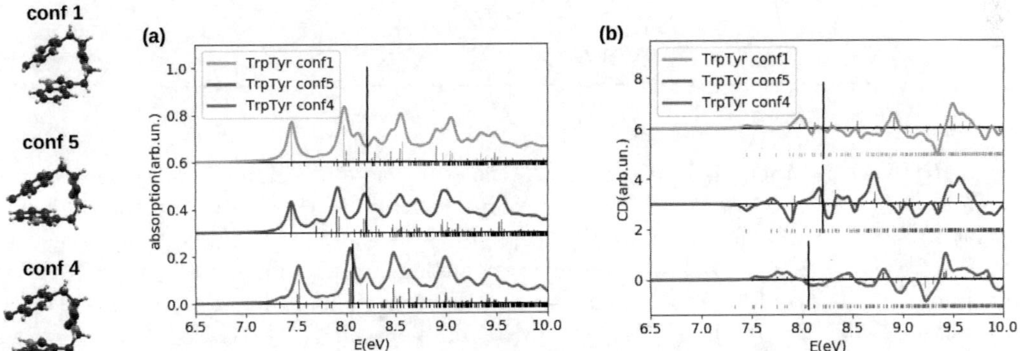

Figure 4. Left panel: geometry of the chosen low-energy conformers of c-TrpTyr. Panel (a): Independent Particle (IP) B3LYP absorption spectra of c-TrpTyr conformer 1 (magenta), conformer 5 (dark red) and conformer 4 (green). Panel (b): IP circular dichroism (CD) spectra of the same three conformers (same color codes). Vertical black lines indicate E(vacuum level) - E(HOMO) for the three conformations. A shift of +2.5 eV has been applied to absorption and CD spectra, and a broadening of 0.05 eV was used.

IOP Conf. Series: Materials Science and Engineering 1265 (2022) 012005 doi:10.1088/1757-899X/1265/1/012005

The mentioned transitions are the only ones giving a non-negligible contribution to IP absorption out of all the transitions up to 8.20 eV. Regarding the spatial localization of the electronic states, as obtained by QE DFT B3LYP calculations in our previous work[4], for c-TrpTyr conformation 1 most of the states near the energy gap, *i.e.* those involved on low energy IP transitions, are localized on a specific part of the cyclic dipeptide, not spread all over the molecule, at a difference with the case of c-GlyPhe conformation 1. In particular, the two transitions giving the most intense contributions to absorption up to 8.20 eV are the HOMO-LUMO and the (HOMO-1) - LUMO one, both involving pairs of states which are localized on the same part of the dipeptide, namely on the indole ring of tryptophan. On the other hand, transitions between pairs of states with different spatial localization yield contributions with lower intensity, such as the above mentioned (HOMO-2) - LUMO and (HOMO-2) - (LUMO+1), or even negligible intensity, such as the transitions from the HOMO state, localized on the Trp indole ring, to states ranging from (LUMO+1) to (LUMO+4), with negligible electronic density on that part of the molecule.

Table 2. Energies (shifted by +2.5 eV), identity of the involved valence to conduction transition, and contribution to absorption and CD for low energy transitions in conformer 1 of c-TrpTyr: asterisks indicate transitions contributing to absorption (column 3) or CD (column 4) peaks.

c-TrpTyr conf1 E(eV)	transition	abs	CD
7.46	H - L	*	* (low)
7.58	H - (L+1)	* (very low)	* (very low)
7.75	H - (L+2)	* (low)	
7.88	H - (L+3)		
7.91	H - (L+4)	* (low)	*
7.98	(H-1) - L	*	*
8.00	(H-2) - L	*	
8.03	H - (L+5)		*
8.10	(H-1) - (L+1)	* (low)	*
8.13	(H-2) - (L+1)	*	*

As for CD spectra, their mutual differences are so pronounced that no recognizable peaks common to the three conformers are present, except possibly for the features in the range from $\approx$ 9 eV to $\approx$ 10 eV. Also for the c-TrpTyr dipeptide the stronger conformational sensitivity of CD spectra with respect to absorption ones is thus confirmed. The first feature in the CD spectrum of c-TrpTyr conformation 1 (magenta curve, right panel in Fig.4) is made of two very weak positive peaks, due to the HOMO - LUMO and to the HOMO - (LUMO+1) transitions respectively (fourth column of Table 2). The (HOMO-1) - LUMO transition yields the most intense contribution to the second (in energy order) CD feature, also of positive sign and more intense with respect to the previous one, lying at the same energies as the second absorption feature; however, this second CD feature has contributions also from two transitions which give low to negligible contributions to absorption, namely the HOMO - (LUMO+4) and the HOMO - (LUMO+5) ones.

For c-TrpTrp the differences among absorption spectra of the three B3LYP lowest energy conformers (see Fig. 5, panel (a)) are larger than those observed for the c-TrpTyr dipeptide, likely due to the larger conformational variability among low energy geometries of c-TrpTrp (left panel of Fig. 5) with respect to c-TrpTyr. In particular, the absorption spectrum of conformer

3 (the B3LYP lowest energy conformer, green curve in panel (a) of Fig.5) is quite different from the spectra of the other two low energy conformers, in the relative intensity of spectral features up to ≈ 8.3 eV, and also, remarkably, in that both the feature at ≈ 7.6 eV and the one at ≈ 8.2 eV consist here of a single intense peak originated by two almost energy-degenerate transitions, with negligible intensity for all the other transitions in that energy range, at a difference with the other two c-TrpTrp conformers, where each of these absorption features originates from several transitions of non-negligible intensity. Interestingly, conformer 3 is rather different from the other two low-energy conformers also in its geometry, suggesting a CH-π interaction for conformer 3, rather than a π-π one as for conformers 1 and 2, as discussed by some of us in our previous work on these cyclo-dipeptides[4].

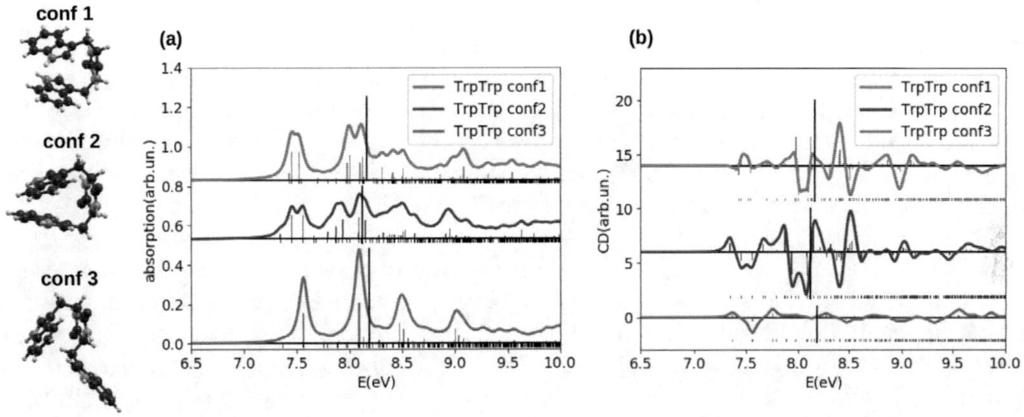

Figure 5. Left panel: geometry of the chosen low-energy conformers of c-TrpTrp. Panel (a): Independent Particle (IP) B3LYP absorption spectra of c-TrpTrp conformer 1 (red), conformer 2 (blue) and conformer 3 (green). Panel (b): IP circular dichroism (CD) spectra of the same three conformers (same color codes). Vertical black lines indicate E(vacuum level) - E(HOMO) for the three conformations. A shift of +2.5 eV has been applied to absorption and CD spectra, and a broadening of 0.05 eV was used.

If we analyze the first absorption feature of c-TrpTrp conf3 (green curve in Fig. 5 panel (a)) in terms of transitions between Kohn-Sham states (short vertical black lines below the spectrum), we find that only the second and the third transition (in order of energy) yield a non-negligible intensity in the energy range up to ≈ 7.8 eV: they are the HOMO - (LUMO+1) and the (HOMO-1) - LUMO transitions (the two intense green vertical lines at ≈ 7.564 eV and at ≈ 7.567 eV, respectively, appearing as a single line in the Figure). Interestingly, the computed B3LYP wavefunctions[4] of the HOMO and LUMO+1 states are both localized on the same part of the TrpTrp dipeptide, *i.e.* on the indole ring of one (the same in both cases) of the two Trp amino acids. The wavefunctions of the HOMO-1 and LUMO states, on the other hand, are both localized on the indole ring of the other Trp. The other transitions in the same energy range, such as the first one in order of energy, HOMO - LUMO, the fourth one, (HOMO-1) - (LUMO+1), the fifth one, HOMO - (LUMO+2), yield negligible intensities, and they correspond to pairs of electronic states localized in different regions of the molecule. The (HOMO-2) - (LUMO+1) and (HOMO-3) - LUMO transitions, both at ≈ 8.09 eV, build the second intense absorption peak for c-TrpTrp conformer 3: they are, once again, transitions between pairs of electronic states localized on the same part of the molecule, *i.e.* the indole ring of either of the two Trp amino acids. As for circular dichroism (Fig. 5 panel (b)), the spectra of conformers 1 and 2 of c-TrpTrp display a similar feature around 8 eV; the spectrum of conformer 3 (the lowest energy geometry) is overall less intense than the spectra of the other 2 investigated conformers. This latter CD spectrum (green curve in panel (b)) displays a first weak positive peak at 7.37 eV, due to the optically dark HOMO - LUMO transition. The following CD peak is more intense, of negative sign, and due to the HOMO - (LUMO+1) and (HOMO-1) - LUMO transitions, *i.e.* the ones involved in the first absorption peak. Then, another positive CD peak is due to the (HOMO-1) - (LUMO+1) transition, which is dark in the absorption spectrum. Finally, the (HOMO-2) - (LUMO+1) and (HOMO-3) - LUMO transitions yield a negligible contribution to the CD spectrum, at a difference with the absorption spectrum, where they yield the second (in energy order) intense peak observed.

4. Conclusion

In this work we have reported on our implementation of circular dichroism calculations at Independent Particle level in the Yambo code and on its application to three cyclo-dipeptides, cyclo(Glycine-Phenylalanine), cyclo(Tryptophan-Tyrosine) and cyclo(Tryptophan-Tryptophan), with some considerations on the more pronounced conformational sensitivity of CD with respect to absorption, and on the interpretation of Independent Particle spectral features in terms of both the energy position of occupied and empty frontier orbitals, and of their spatial localization. The implementation of CD calculations beyond IP level will be the subject of future works.

In particular, an analysis of Independent Particle absorption spectra of the three investigated cyclo-dipeptides, together with the spatial localization of the electronic states involved in the optical transitions (see Results section) illustrates how similar densities of electronic energy levels can potentially yield rather different absorption spectra, due to the generally higher (lower, respectively) intensities of contributions from transitions between pairs of orbitals with a large (resp. small) spatial overlap. This can happen, *e.g.*, for different conformers of the same molecule, when considering transitions between pairs of orbitals which are localized on parts of the molecule that are close to each other in one conformer and at a larger distance in another one. We can expect such effects when the molecule is not too small, it has a large conformational variability and the involved orbitals are not spread all over the molecule. A good candidate, in the present work, is the c-TrpTrp molecule, whose lowest energy conformer has a set of frontier orbitals which are alternately localized on the indole ring of either of the two tryptophan amino acids, as shown by some of us in a previous work[4]). Indeed, as we have discussed above (Results section), Independent Particle absorption spectra of the three lowest energy conformers of c-TrpTrp (Figure 5) are rather different from each other. Moreover, the fact that the shape of circular dichroism spectra is affected by the positive or negative sign of the CD contributions corresponding to individual optical transitions, in addition to their intensity, and that the relative intensities of individual contributions to CD are not necessarily proportional to the corresponding contributions to absorption, increases the conformational variability of CD spectra with respect to absorption ones. This renders CD spectroscopy potentially useful for conformational analysis, while at the same time requiring careful consideration when using it for identifying enantiomers of flexibile molecules.

Acknowledgments

The present work was performed in the framework of the PRIN 20173B72NB research project "Predicting and controlling the fate of biomolecules driven by extreme-ultraviolet radiation", which involves a combined experimental and theoretical study of electron dynamics in biomolecules with attosecond resolution. G.C. acknowledges collaboration with "Progetti biennali d'Ateneo UniCa finanziati dalla Fondazione di Sardegna annualità 2020: Molecular simulations and machine learning in bio- and medical physics: from antimicrobials to radiopharmaceuticals", and partial financial support from IDEA-AISBL-Bruxelles. We moreover acknowledge financial support from European Union project MaX Materials design at the eXascale H2020-EINFRA-2015-1 (Grant Agreement No. 824143).

References

[1] Molteni E, Onida G, Ceccarelli M and Cappellini G 2021 *Symmetry* **13** 601
[2] Molteni E, Onida G and Tiana G 2015 *J. Phys. Chem. B* **119** 4803–4811
[3] Sangalli D *et al.* 2019 *Journal of Physics: Condensed Matter* **31** 325902
[4] Molteni E *et al.* 2021 *Phys. Chem. Chem. Phys.* **23** 26793–26805
[5] Mishra A K, Choi J, Choi S J and Baek K H 2017 *Molecules* **22** 1796
[6] Zhao K, Xing R and Yan X 2020 *Peptide Science* e24202
[7] Jeziorna A, Stopczyk K, Skorupska E, Luberda-Durnas K, Oszajca M, Lasocha W, Górecki M, Frelek J and Potrzebowski M J 2015 *Crystal Growth and Design* **15** 5138–5148
[8] Mattioli G *et al.* 2020 *Sci. Rep.* **10** 13081
[9] Ying J, Lin R, Xu P, Wu Y, Liu Y and Zhao Y 2018 *Scientific Reports* **8** 936
[10] Danger G, Plasson R and Pascal R 2012 *Chem. Soc. Rev.* **41** 5416–5429
[11] Barreiro-Lage D, Bolognesi P, Charinelli J, Richter R, Zettergren H, Stockett M H, Carlini L, Diaz-Tendero S and Avaldi L 2021 *J. Phys. Chem. Lett.* **12** 7379–7386
[12] Giannozzi P *et al.* 2017 *J. Phys. Cond. Mat.* **29** 465901
[13] Giannozzi P *et al.* 2009 *J. Phys. Cond. Mat.* **21** 395502
[14] Troullier N and Martins J L 1991 *Phys. Rev. B: Condens. Matter Mater. Phys.* **43** 1993
[15] Varsano D, Espinosa-Leal L A, Andrade X, Marques M A L, di Felice R and Rubio A 2009 *Phys. Chem. Chem. Phys.* **11** 4481–4489
[16] Becke A D 1993 *J. Chem. Phys.* **98** 5648
[17] Lee C T, Yang W T and Parr R G 1988 *Phys. Rev. B: Condens. Matter Mater. Phys.* **37** 785
[18] Grimme S, Antony J, Ehrlich S and Krieg H 2010 *J. Chem. Phys.* **132** 154104
[19] Makov G and Payne M C 1995 *Phys. Rev. B* **51** 4014–4022
[20] Condon E U 1937 *Rev. Mod. Phys.* **9** 432–457
[21] Barron L 2004 *Molecular Light Scattering and Optical Activity* (Cambridge: Cambridge University Press)
[22] Carnell M, Peyerimhoff S D, Breest A, Goedderz K H, Ochmann P and Hormes J 1991 *Chem. Phys. Lett.* **180** 477–481
[23] Catalan J 2016 *Phys. Chem. Chem. Phys.* **18** 15170–15176
[24] Hazra C, Samanta T and Mahalingam V 2014 *J. Mater. Chem. C* **2** 10157–10163
[25] Tao K *et al.* 2018 *Nat. Comm.* **9** 3217

Different Aspect Ratio Polysaccharide-Assisted Assembling of Magnetic Nanoparticles

Riccardo Di Corato[1,2,6,*], Alessandra Aloisi[1,6], Giacomo Mandriota[3,7], Anna Penna[4], Laura M. Slavu[3] and Rosaria Rinaldi[1,3,5,*]

[1]Institute for Microelectronics and Microsystems (IMM), CNR, Via Monteroni, Lecce 73100, Italy

[2]Center for Biomolecular Nanotechnologies, Istituto Italiano di Tecnologia, Arnesano 73010, Italy

[3]Mathematics and Physics "E. De Giorgi" Department, University of Salento, Via Monteroni, Lecce 73100, Italy

[4]ICONplc, Via Crespi 19, 20159 Milano. Corporate Headquarters: South County Business Park, Leopardstown, Dublin 18, Ireland

[5]Scuola Superiore ISUFI, University of Salento, Via Monteroni, University Campus, 73100 Lecce, Italy

[6]Dr. R. Di Corato and Dr. Alessandra Aloisi contributed equally to this work.

[7]Current affiliation: Istituto per i Processi chimico Fisici (IPCF), CNR, Via Orabona 4, Bari 70126, Italy.

*Corresponding authors e-mail: riccardo.dicorato@cnr.it, ross.rinaldi@unisalento.it

Abstract. Versatile methods are here presented for the assembling of magnetic nanoparticles in controlled super-structures with different final aspect ratios. The first step of the procedure is based on the destabilization, and consequently clustering, of a dextran-coated nanoparticles suspension, altering the colloidal stability of the initial building blocks. Subsequently, a second polysaccharide shell, based on chitosan, is grafted to the super-structure surface, for stopping the aggregation of the nanoparticles and enhancing the stability of the resulting assembly. For achieving a finer control of the process, spherical magnetic clusters (SMC) are prepared through a microfluidic aided clustering, investigating the contribution of flow rates and injection scheme geometry to the nanoparticle assembly. Besides, the use of static magnetic fields applied during the first step results in the preparation of micrometric elongated magnetic super-structures (EMS), tuning the average aspect ratio up to a value of 13±4.

1. Introduction

In the massive field of nanotechnologies, colloidal nanoparticles have increasingly found several uses in medicine (resulting in the emerging topic of nano-medicine) and biomedical applications. Within this context, magnetic nanoparticles are considered one of the main interesting materials, with a solid clinical application as MRI contrast agents[1] and an extensive commercial usage as tools for magnetic separation of analytes [2]. Furthermore, research on the exploitation of these nanoparticles as diagnostic (e.g. MRI, Magnetic Particle Imaging (MPI)) or therapeutic (e.g. magnetic hyperthermia, photoablation, drug delivery, tissue engineering) agents aims towards a future use in the medicine, especially when incorporated in multifunctional systems [3]. Another aspect related to magnetic nanoparticles is the controlled assembly in ordinated super-structures. This topic arose a particular interest in last years, resulting in the preparation of several kinds of multifunctional systems with peculiar properties deriving from the interaction between single nanoparticles in a dense core. Indeed, when the individual crystals are confined in a superstructure, specific forces are enhanced or triggered, bringing to new functions [4].

In the case of magnetic nanoparticles, the superparamagnetic profile, which is peculiar to nanoparticles with a diameter smaller than a certain threshold (about 20 nm for iron oxide), is also maintained for the superstructures obtained through the controlled assembly of single nanoparticles. Moreover, phenomena of increase of other magnetic properties, such as R_2 relaxivity or saturation magnetization, have also been observed, due to the enhanced inter-particles interactions inside the magnetic core. Surely, one of the main advantages of a controlled superstructure is the possibility to manipulate it macroscopically by external magnetic fields. In fact, in a dilute system (such as the bloodstream) the low concentration of nanoparticles dramatically limits the magnetic response. Instead, the cumulative effect deriving from the increased concentration of the crystals packed into the magnetic cores of the nanostructures allows the single multidomain to be attracted and manipulated even in diluted systems. On the other hand, the increase in size from a single magnetic core (3-20 nm) to a superstructure (> 60-80 nm) sometimes compromises the colloidal stability of the structure, which use must therefore be well-calibrated to avoid severe adverse effect. In the literature there are many different references concerning quasi-spherical magnetic superstructures [5-9]; in addition, also some elongated structures have been reported (i.e. in particular magnetic nanowires, nanorods, chain, etc.) specifically used in different applications, exploiting their potential in cell manipulation, in microfluidic systems, micromechanics, microreology [10-11]. A general strategy for obtaining such nano or micro-structures is based on the phenomenon of the co-assembly of polymeric microspheres of nanoparticles, generating chains, thanks to the presence of an external magnetic field and the addition of specific ligands. The use of a permanent magnet is the essential key to guide the assembly of superstructures in 3D space, generating an attractive force capable of aligning the particles with each other in the resulting structure, in the desired pattern.

In the design of nano-systems, both in the case of single nanocrystals and superstructures, the choice of the most appropriate coating agent is of crucial importance. This will determine the colloidal stability, the interaction with the biological elements and the possibility of using the system as cargo for certain payloads. One of the most used molecules for the stabilization of magnetic nanoparticles is represented by polysaccharides. These molecules are often obtained by a natural source and possess innate biocompatibility. It is interesting to note that the first FDA-approved formulations based on magnetic nanoparticles (*e.g.* ferucarbotran and ferumoxide) were obtained by aqueous assisted nucleation of magnetic seeds in the presence of a modified-dextran [12], although some of the above-mentioned formulations have been stopped for further development or withdrawn from the market [13]. Starting from hydrophobic nanoparticles, polysaccharides have been widely used for the stable dispersion of single cores [14-16] but also for the preparation of small nanoclusters ranging from 100 to 400 nm [17-18].

This contribution aims to investigate different approaches for the destabilization of dextran-coated superparamagnetic nanoparticles (D-MNP), obtained by the well-known coprecipitation method [19]. The strategic innovation is to modify, in a controlled way and through the use of organic solvents, the colloidal stability of the obtained nanoparticles, favoring the formation of spherical magnetic clusters (SMC) of nanoparticles, which are potentially more responsive to the presence of an external magnetic field than single nanoparticles. A microfluidic-assisted destabilization and rearrangement approach is also proposed, to improve control over the size, size distributions, and compositions of the described structures.

In a second step, the introduction of an external magnetic field during destabilization allows changing the final structure of the magnetic material, achieving a stable elongated magnetic superstructure (EMS).

Remarkably, the aptitude of low molecular weight chitosan (Cs) to participate in polyelectrolyte complexes (PECs) by simple coacervation with negatively charged polymers (i.e. dextran, alginate, carboxymethyl cellulose, pectin, carrageenans, and other polyanions) [20] is also here exploited. The dextran is a branched polysaccharide of variable molecular weight, produced by lactobacilli or other fermenting bacteria from sucrose by enzymatic activity. It is composed of a succession of D-glucose molecules linked by an α1-6 glycosidic bond with short branches in the α1-3 position. The numerous available hydroxyl groups anchor robustly to the surface of iron oxide nanoparticles, enhancing the long-term stability of the starting magnetic particle preparation.

Resulting from the partial N-deacetylation of chitin, Cs is a linear polysaccharide composed of randomly distributed β-(1-4)-linked D-glucosamine and N-acetyl-D-glucosamine. The key structural parameter of chitosan is the degree of acetylation (DA), which is responsible for both the charge density of the polymer and the balance between the hydrophilic and hydrophobic interactions.

PECs based on chitosan as polycation and dextran as polyanion are largely prepared by simple coacervation. The mode of addition, charge mixing ratio, ionic strength and pH of the media, and the molar mass of both components have been exhaustively reviewed by Schatz et al. [21].

Modulating Cs concentration, it becomes possible to fine control super-assembled structure in terms of compactness and stability; to avoid aggregation of the PEC structures, the polyelectrolyte solutions must be diluted, and one of the polyions must be in the appropriate excess so that the charge ratio $(n+/n-) \neq 1$ [22-23]. The arrangement of chains in a PEC can be pictured as a combination of a disordered scrambled egg-like structure and a highly ordered ladder-like organization. Therefore, the actual structure having hydrophobic and hydrophilic regions makes PECs a particular class of physically cross-linked hydrogels that are sensitive to environmental factors [20].

Low-molecular Cs degradation products, the chito-oligomers, result mainly from the influence of enzymes, which are found in the organism fluids. Actually, computational predictions revealed that investigated small chito-oligomers, regardless of molecular weight, acetylation degree and acetylation pattern, reflect favorable pharmacological profiles [24]. Several agents have been proposed for additional ionic or covalent crosslinking of Cs available amino groups, allowing to control coating degradability for diverse applications [25-26]. Moreover, assuming future exploitation of these microstructures in nanomedicine, the chitosan immune biocompatibility and low immunomodulating properties [27] could promise an advantage for *in vivo* nanomedicine application of Cs encapsulated functional superstructures.

2. Results

The synthesis of the superparamagnetic nanoparticles was carried out by using a well-established coprecipitation method in the presence of dextran molecules. In this case, the obtained crystals are very similar to the standard Ferucarbotran nanoparticles, in which, the precipitation of iron oxide in an alkaline solution is modulated by the presence of the polysaccharide, that figures as a template for blocking and preventing the uncontrolled aggregation of the superparamagnetic particles.

2.1. Destabilization in tube

The destabilization of the system, also known as clustering, was achieved by intervening on the stability of the nanoparticles suspension (and therefore on the corresponding dextran outer layer) by means of the controlled addition of an organic solvent. The choice of the solvent was performed following some preliminary experimental observations with reference to parameters of polysaccharide solubility, polarity and miscibility. The effects of some destabilizing solvents including ethanol, acetonitrile, acetone and dimethylformamide on D-MNP were summarized and commented in Table A1. The experimental observations pointed out acetonitrile (ACN) as the most suitable destabilizing solvent to reach the final purpose. All the experiments were conducted in 5 main steps: i) injection of the destabilizing agent in a suspension of D-MNP (destabilization step); ii) qualitative selection of samples based on the suspension scattering; iii) chitosan coating of the superstructures (stabilization step); iv) separation / purification to the magnet; v) morphological analysis.

One of the first optimized parameters consisted in the acetonitrile addition rate. A flat rate was not optimal in terms of morphology and distribution. A multistep injection was preferred, with a faster injection in the first step (100 $\mu L\ min^{-1}$ for 3 minutes), slowing down then the rate to 25 $\mu L\ min^{-1}$ for the remaining volume. Once fixed the rate flow scheme, a series of experiments were carried out by varying the amount of injected acetonitrile, to determine an optimal percentage that avoids a disruptive aggregation of the system (Table A2, lines 1-3). It is noteworthy that at least 40% of acetonitrile is necessary to destabilize the system; working with a lower percentage, no clustering was detected. However, the clusters were still inhomogeneous and irregular in shape (Figure A3). The obtained SMC were stabilized immediately adding a solution of chitosan, dispersed in a mixture of water and ACN, to obtain a final polysaccharide concentration of 1%.

To achieve better control on the clustering, another important parameter was the amount of initial nanoparticle involved in the destabilization. Lowering the volume to 10 and 5 μL (Table A2, lines 4-5), spherical superstructures were obtained, with an average size of 300 nm (Figure 1). However, the size was still very polydisperse (ranging from 100 to 500 nm) and the nanoparticles assembly in the core was not dense enough. These results encouraged the evaluation of another strategy of clustering, shifting from a bulk reaction in tube to a more controlled aggregation in the microfluidic channel.

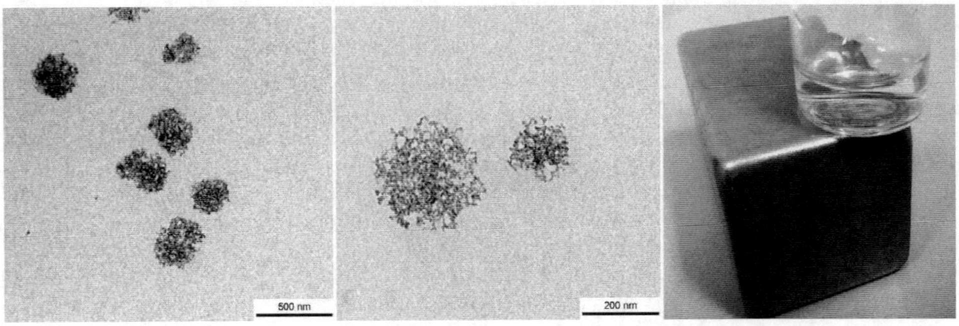

Figure 1. TEM images of SMC obtained by in tube destabilization (Table A2, sample 4). On the right, a photograph of SMC suspension magnetically sorted in few seconds.

2.2. Destabilization in device

The best conditions for the destabilization process, selected above in the tube reaction (section 2.1), have been translated into the microfluidic device, in order to obtain a better control of the destabilizing agent injection process and the consequent formation of smaller and more compact aggregates (SMC). The dextran-coated magnetic nanoparticles were destabilized with ACN through a system consisting of the microstructured PDMS/glass device, two syringe pumps and an orbital shaker for collecting the obtained superstructures under continuous stirring. This approach is aimed to evaluate the possible effect that fluid-dynamic processes at the interface (occurring at the microscale) may have on the control of particles destabilization and rearrangement.

As a preliminary set of tests, the variations of the concentrations of the D-MNP and the percentages of ACN (obtained by complementary infusion rates for an overall rate equal to 1 mL min^{-1}) were evaluated (Table A3, samples 1-5). For each of the different five tests, 1 mL of pure acetonitrile was injected through the main inlet and a suspension of D-MNP (2.5 mM, in ultrapure water) was pumped from the two lateral channels, orthogonal to the main inlet.

The clusters were collected in a vial (on an orbital shaker) and then stabilized by adding a solution of Cs (1% in 38% ACN / 62% water), to obtain a final Cs concentration of 0.5%, and allowed to interact for 2 hours before magnetic sorting and purification.

By TEM characterization, the samples 1 and 2 showed the formation of well-defined spherical nanoclusters, with an average size from 150 nm to 500 nm (sample 2 in Figure 2). However, the low percentage of chitosan in the stabilization step was not sufficient to obtain a suitable shield/stabilization of the superstructures during the washing steps to the magnet. Experiments 3, 4 and 5 showed no evident formation of magnetic nanoclusters.

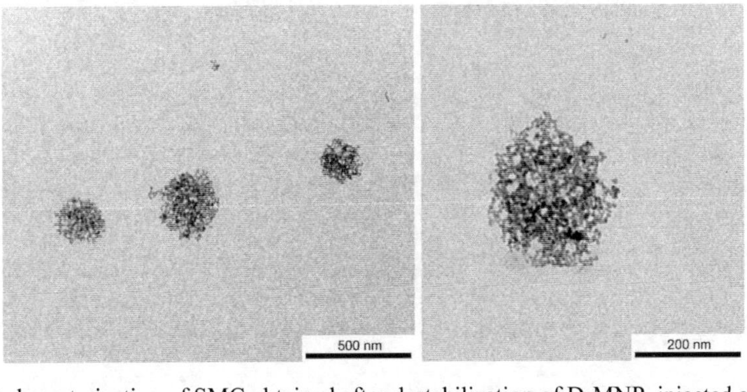

Figure 2. TEM characterization of SMC obtained after destabilization of D-MNP, injected at a rate of 0.275 ml min^{-1}, with ACN injected at a rate of 0.45 ml min^{-1} and subsequent stabilization in Cs 0.5% / ACN 38% (final concentrations, Table A3, sample 2).

Taking into account the preliminary experiments (samples 1-5, Table A3), in samples 6-8 (Table A3) the destabilization was reproduced by injecting 1 ml of ACN from the main inlet at a flow rate of 0.45 mL min^{-1} and a suspension of D-MNP ([Fe] = 2.5 mM) from side arms at a rate of 0.275 mL min^{-1}. The overall collected volume of destabilized nanoparticles is 1.6 mL. As a control of the established protocol, an aliquot of freshly collected SMC was taken, drop-casted on a copper grid and left to dry in an ACN saturated chamber for TEM characterization (Table A3 line 6, Figure A4). The solvent saturated chamber drastically increased the evaporation time of the sample drop, preserving the morphology of the obtained superstructures; the uncoated sample cannot be subjected to magnetic separation as the 3D structure, although well-formed, tends to disassemble in this process, if not further stabilized. The obtained clusters were spherical, very dense and with an average size of 500 nm, thus confirming the results obtained in the previous step. The nude nanoclusters were then stabilized by adding a solution of Cs (in a mixture ACN / water), in order to obtain a final concentration of 1%. The percentage of acetonitrile (45%) was raised by pairing the same resulting in the destabilized collected volume (Table A3 line 7, Figure A5). The obtained cluster showed a smaller size, still with a dense core but not completely packed anymore.

Having ascertained that the Cs coating is of considerable importance for structural stability during the SMC separation and purification step to the magnet, as well as for the stability in the aqueous phase for subsequent bio applications, for the latter experiment we proceeded to optimize the stabilization process, maintaining the same polysaccharide concentration but increasing the percentage of ACN by 5%, bringing it to 50% (Table A3 line 8, Figure 3). By TEM characterization, the SMC were quite regular in shape and homogeneous; the density of magnetic nanoparticles was preserved in comparison to the naked cluster, with a narrow size distribution between 325 and 450 nm.

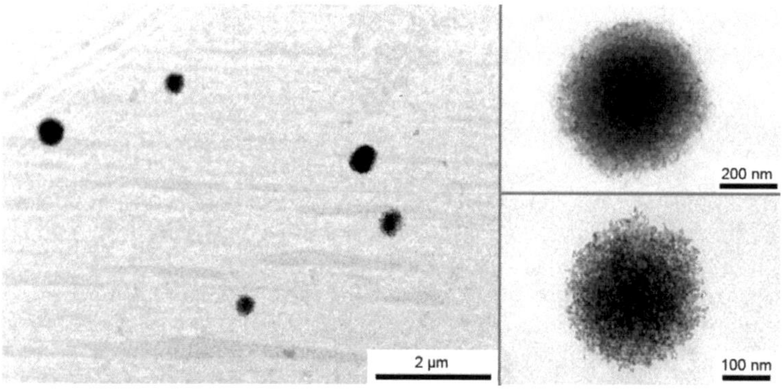

Figure 3. TEM images of characterization of SMC (Table A3, sample 8). SMC obtained after destabilization of D-MNP, injected at a rate of 0.275 ml min^{-1}, with acetonitrile injected at a rate of 0.45 ml min^{-1} and subsequent stabilization in Cs 1% / ACN 50% (final concentrations).

The SMC were analyzed by DLS, to evaluate the dispersity and the stability of the superstructures. The hydrodynamic diameters of dextran-coated nanoparticles and SMC were compared (Figure A6A); for the single particles an average diameter of 46 ± 18 nm was detected, with a ζ-potential of +2.3 mV. Analyzing the SMC obtained in the experiment reported in Figure 3 an increase in diameter was detected, with an average value of 193 ± 46 nm (PDI = 0.1). Because of the Cs outer layer, the ζ-potential of SMC was strongly positive, with a value of +53 mV. The same value was measured for SMC destabilized in tube (section 3.1). The colloidal stability of the SMC was evaluated by monitoring the measured mean count rate. The SMC were dispersed in DMEM culture medium and analyzed every two minutes in a 60 minutes time lapse (Figure A6B). The graph showed fair stability over time of SMC, whit an initial 5% decrease of count rate in the first 10 minutes (probably due to some bigger clusters precipitation) and an acceptable 10% decrease in the remaining 50 minutes.

The performed tests were aimed at optimizing the experimental parameters, to obtain a more compact, smaller and stable SMC, controlling the destabilization process using a microfluidic approach. From the first trials (samples 1-5), a better destabilization in terms of turbidity was observed compared to the in-tube tests. Then, the increase of the final Cs percentage (1%) allowed a better stabilization and packing, as the increase of ACN environment (50%) applied during the last step (out of the device). Moreover, the use of a syringe pump-controlled injection system, in the microfluidic device, helped to have better control over the flow rate and the generated flow pattern. To document the flow dynamics inside the device, Figure 4 shows a demonstrative test of what happens in the microscale system, at the concentration and rate conditions optimized for the destabilization of the dextran-coated magnetic nanoparticles.

Flow focusing consists in injecting a core liquid into another surrounding flowing sheath liquid. Here we exploited the influence of injection imposing pressure to generate a parallel 3-stream pattern of the two miscible liquids at stake and the consequent inertial phenomena. We injected ACN (Dyn. Viscosity 0.334 cP, 25°C) in the central inlet of the cross-junction microfluidic device and a water (Dyn. Viscosity 0.89 cP, 25°C) suspension of D-MNP in the two lateral inlets. The side/core flow rate ratio allowing the best results corresponds to a value of 1.2. (Table A3). Mainly, three forces can be considered as account for droplet/D-MNP rearrangement dynamics in inertial focusing flows: a wall interaction force, a shear gradient lift force, and a secondary-flow drag force as discoursed in [28].

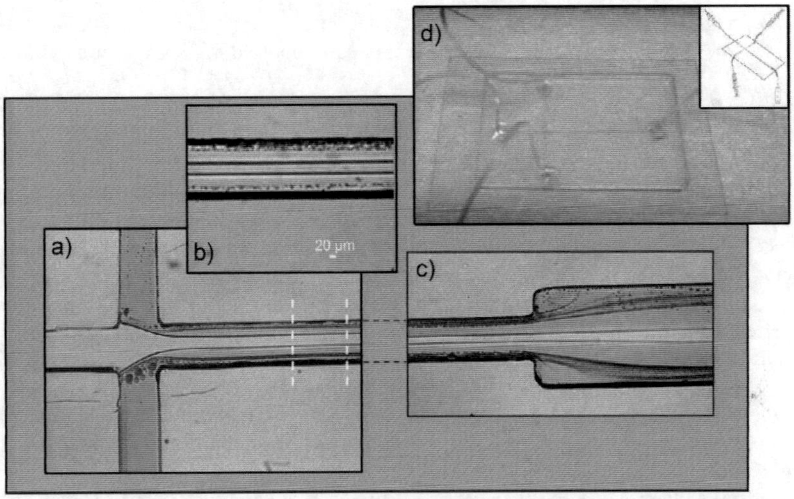

Figure 4. a-c) Brightfield photomicrograph (obj. 4x). Detail of two different sections of the microchannel system. On the left (panel a), the cross junction near the inlet region where the laminar flow generated by the injection is observed at V = 0.45 ml min^{-1} of ACN and V = 0.275 ml min^{-1} of aqueous solution (orange-colored to facilitate the observer); on the right (panel c), the final stretch near the outlet region, where a turbulent motion is generated (at zero pressure). Note, along the main microchannel (highlighted in panel b) the generation of microdroplets (in the H_2O / ACN mixture) mainly occurring on the PDMS hydrophobic surface. The continuous droplets moving is assumed to participate in a phenomenon of NPs squeezing in the process of destabilization. In panel d, a photograph and a sketch of the device are shown.

2.3. Destabilization in presence of magnetic field: Elongated Magnetic Superstructures (EMS)

The principles of destabilization previously described were exploited to evaluate the effect of the introduction of an external magnetic field applied during the destabilization phase (and in the subsequent steps), aimed at obtaining asymmetric macrostructures with an elongated morphology. The geometries applied in the various working hypotheses were chosen to evaluate the influence of the static field; in a first analysis it is possible to state that the positioning of the magnet and the strength of the field were found to be decisive for obtaining satisfactory results. The three different permanent magnets (Circular 1 – C1, Circular 2 – C2, Parallelepiped – P, Figure 5) were physically positioned near the glass vial containing the aqueous suspension of D-MNP, at a fixed distance of 7.5 mm (from the outer edge) placed under agitation on an orbital shaker for tubes or plates.

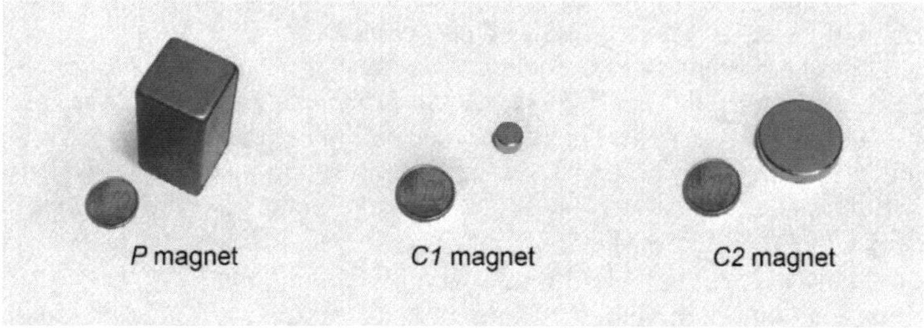

Figure 5. Photograph of the magnets used for the EMS preparation. Magnet *P*: nominal field 0.4 T, size 50x25x25 mm. Magnet *C1*: nominal field 0.1 T, size 10x5 mm. Magnet *C2*: nominal field 0.2 T, size 30x5 mm. Magnetic gradients analyses of each magnet are reported in the experimental section.

The addition of the destabilizing solvent, ACN, was carried out manually, following the same flow rate scheme adopted for SMC in tube preparation. It has been observed that the addition of the first 300 μL of solvent does not create substantial changes in the sample under examination, while the addition of the last 100 μL leads to the reaching of the saturating limit for which the magnetic nanoparticles precipitate forming aggregates that are affected by the applied magnetic field going to arrange and align in elongated superstructures. The occurred destabilization is evaluated macroscopically thanks to the slight turbidity of the solution. After ACN addition, the suspension was left to stir for 1 minute, 5 minutes or 15 minutes, depending on the tests, allowing the system to create the right interaction between the applied magnetic field and the destabilized material and then proceed with the drop-by-drop addition of chitosan for blocking the controlled aggregation and stabilizing the obtained EMS.

2.3.1. C1 magnet configuration. First tests were carried out with C1 magnets, assembled in two or four units and positioned on the sides of the glass vial containing the aqueous solution of nanoparticles. The distance between the two opposite magnets has been set at 32mm (Table A4).

It has been observed that by leaving the aggregates to stir in the presence of the magnet for more than 10 minutes, before adding the Cs, a total aggregation of the system under examination was obtained. For this reason, it was appropriate to introduce the polysaccharidic coating at a time between 1 and 5 minutes after ACN addition. The TEM images in Figure 6A show the first obtained results, applying four magnets per side. The formation of hemispherical aggregates is highlighted, slightly different from the destabilized material obtained in the absence of an external field. All the tests listed in Table A4 from B to E were carried out using 2 magnets per side. At first glance, early elongated structures are highlighted. By fixing the concentration of D-MNP at 2.3 mM, parameters such as the ACN added volume and the final Cs concentration were varied. It is noteworthy that in the samples B and C (Figure 6B and 6C) the low stirring speed, respectively 400 rpm and 1000 rpm, was not sufficient to supply the system with the proper energy to favor the magnetic alignment, therefore mostly spherical aggregates were obtained. The results shown in figure 6D, for sample D, disclosed small elongated structures in which it is possible to distinguish and identify the single spherical nuclei that tend to blend in the assembly of a single structure. The best results for this series of tests were obtained in sample E (Figure 6E), in which more defined and dense structures, of elongated shape, but not yet perfectly delineated, were observed. It is possible to state that the strength of the field is not such as to allow an interesting alignment (regardless of the number of coupled magnets and the possibility of placing them on both sides of the reaction vial) and therefore in the following tests larger magnets will be used.

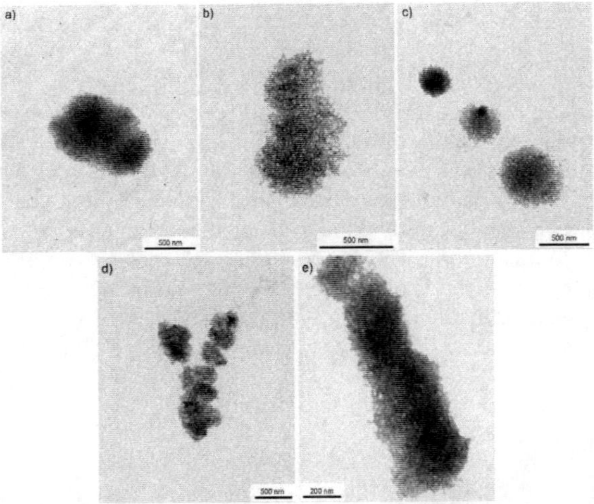

Figure 6. TEM images for the best EMS preparations obtained by C1 configuration, respectively from sample A to sample E. Additional images are reported in Figure A7.

2.3.2. C2 magnet configuration. The tests performed by using the C2 magnets were carried out by assembling them in several units, from 3 to 5 components, evaluating the substantial differences created by the introduction of a new magnetic field with circular geometry, keeping the necessary volume of acetonitrile constant to destabilization. All obtained structures were fixed in chitosan after 5 minutes from the destabilization. In this series of experiments, the most significant results were achieved by using a combination of at least 4 single magnets, and by injecting a volume of acetonitrile equal to 400 μL (Table A5). Comparing the samples A and E (Figure 7 and A8), both obtained through the use of 4 magnets, it can be seen that sample E shows a considerable stretching of the elongated structures under examination, fixed specifically with 2 mL of 2% Cs (initial concentration), unlike for sample A where only 1 mL of Cs was used in stabilization step. Thus, this is a confirmation that a larger volume of chitosan was needed to firmly pack the formed structures. Sample D, opposed to A and E, shows how in this series of experiments a volume of acetonitrile less than 400 μL may not be concretely sufficient to induce the elongated structural organization of the nanoparticles within the system.

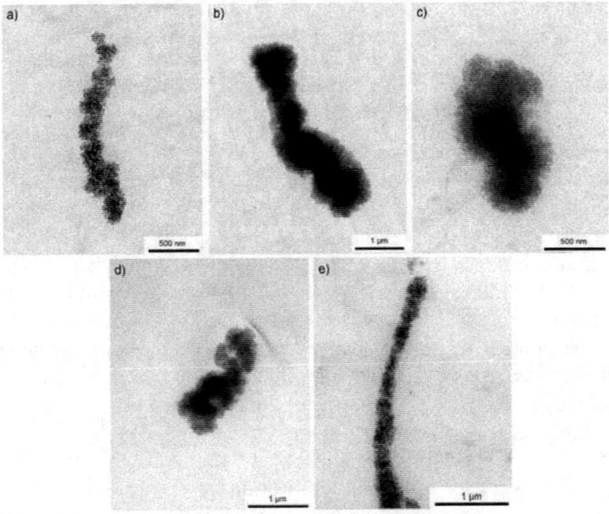

Figure 7. TEM images for the best EMS preparations obtained by C2 configuration, respectively from sample A to sample E. Additional images are reported in Figure A8.

2.3.3. P magnet configuration. For this series of tests (Table A6), the large magnet P was placed in a horizontal position close to the vial. As a direct effect of this setup, the magnetic nanoparticles were more affected by the influence of the external magnetic field applied, going to assemble into much more stretched and elongated structures, and with a quite regular shape. Samples A, B, C (Figure 8 and A9) were obtained by destabilizing the system with a final volume of ACN equal to 350 μL; the structural variations between the 3 tests are limited to the waiting times before the final addition of chitosan, respectively equal to 1 minute (A), 5 minutes (B), 15 minutes (C). The best results are achieved by sample B corresponding to a volume of ACN of 350 μL and the addition of Cs after a waiting time of 5 minutes. Sample D was obtained using a final volume of ACN equal to 400 μL, with chitosan solution added after 5 minutes; although the conditions used were quite similar to sample B, the increase of the ACN volume did not result in a morphology improvement.

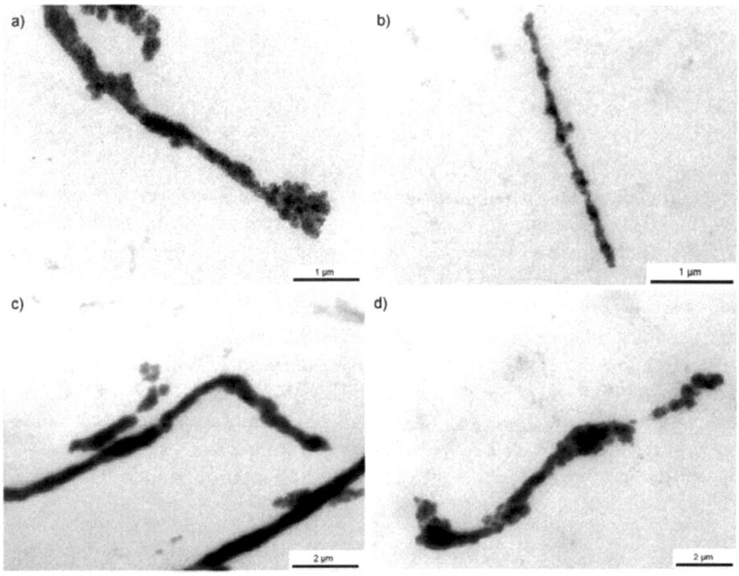

Figure 8. TEM images for the best EMS preparations obtained by P configuration, respectively from sample A to sample D. Additional images are reported in Figure A9.

3. Discussion

The controlled aggregation of superparamagnetic nanoparticles in superstructures, namely clustering, offers the advantage of maintaining the superparamagnetic behavior of the individual nanoparticles, while there is a faster accumulation on the magnet due to the cumulative magnetization effect of different magnetic nanoparticles in the inner core of the assembly. Usually, this type of superstructures can be obtained through a destabilization process of the surface coating, through the use of partially soluble / miscible organic solvents such as to cause an increase in surface tension. If the process is properly guided, an increase in size and a control over the three-dimensional structure are obtained. Starting from dextran-coated magnetic nanoparticles as building blocks, the clustering was induced by controlled addition of a polar aprotic solvent as the ACN. It was found that the variations in volume and flow rate of the destabilizing agent deeply affect the destabilization process of magnetic nanoparticles. Moreover, an improvement in the stability of the superstructures was achieved by diluting the Cs, for the subsequent coating of the SMC, in an ACN solution possessing a percentage slightly higher than that used for the destabilization step.

The interest was then essentially focused on the possible effect that the fluid-dynamic processes at the interface, which occur at the microscale, can have on the control of the destabilization of magnetic nanoparticles. Fluid dynamics inside microchannels is governed by a laminar regime where the capillary force acts on the fluid flow, depending on the surface energy of the channel walls and/or the surface tension of fluid itself. Exploiting 2D hydrodynamic flow focusing on microfluidic channels, the ACN stream was horizontally focused by water sheath streams. After 24 h from plasma bonding, the PDMS native hydrophobicity recovery prompted the generation of microdroplets (in the H_2O / ACN mixture). The continuously moving droplets are assumed to participate in the squeezing of NPs in the process of destabilization and spherical rearrangement.

Our results demonstrated that the microfluidic device could be exploited as an effective tool for controlling NP aggregation process and rearrange magnetic clusters with improved magnetic properties. Although conventional bulk mixing occurs at the timescale of ten second order, in our microfluidic device the mixing time of solvents was controllable and tunable, opening new strategies to synthesize nanomaterials with more precise dimensions at narrow size distribution. Of note, the crucial role exerted by PECs, concerning the stability and the density of packed magnetic clusters.

Focusing on the elongated magnetic superstructures, from the TEM images it can be seen that an average *aspect ratio* value of 13 ± 4 has been obtained in the best tests with more elongated superstructures. It is estimated that these dimensions may be favorable for evaluating the application of EMS in the cell in the presence and absence of a magnetic field. Furthermore, the Cs coating makes the superstructure particularly biocompatible for a biological application.

Analyzing the characteristics relating to the morphology acquired by the EMS, it is possible to think that they may be able to penetrate the cell membrane with greater force, unlike a spherical geometry that requires biological phenomena of internalization and particularly high magnetic fields. The purpose is therefore the damage and alteration of the normal physiology of the cell membrane by creating permanent or transient pores capable of generating an alteration of the homeostatic balance such as to induce a state of cellular necrosis [29-30]. In the same way, damage to essential structures such as cytoskeleton and mitochondria could be evaluated, together with other organelles responsible for key functional roles for cellular integrity.

Another significant application is the use of a dynamic magnetic field, capable of supplying the EMS with the mechanical energy necessary to overcome the elasticity of the cellular target. Even assuming a poor induction of permanent cell damage through the exclusive interaction between EMS and membrane, there is the possibility of exploiting a whirling field to operate single EMS within the cell cytoplasm. These could cause supercoiling of protein structures, as well as the escape of enzymes such as catalase and cytochrome from the mitochondria or damage to the nuclear membrane, irreversibly activating apoptotic processes in target cells [11]. Moreover, the potential magnetic targeting and the asymmetric morphology could minimize the immune clearance of the nanocomposite (overcoming the passive targeting) and boost its tumor accumulation by "destructuring" the interstitial tumor matrix [31].

4. Conclusions

The presented methodology and reported experimental findings focus on different possible approaches for the super-assembly of magnetic nanocrystals, basing the clustering procedure on the controlled destabilization of dextran-coated nanoparticles. By exploiting the different degrees of solubility of the polysaccharide in different solvents, the progressive destabilization of an aqueous suspension of nanocrystals was governed by the addition of ACN. In the destabilization steps performed both in the tube and in the device, the addition rate of the organic solvent was of crucial importance: a very fast addition results in a decreasing control on cluster morphology and a coarse aggregation. Conversely, a too slow addition does not allow to obtain clusters with a good density. The D-MNP destabilization process occurs when the percentage of ACN reaches a critical threshold, equal to 38-40 %, for which the flow control assumes significant importance, through the addition of the solvent by controlled pumping systems. By translating this concept into microfluidics, the reaction starts at the right angle intersection between the central and side channels, promoting over a short time the formation of more regular superstructures. In these experiments, it was found that the ACN flux, and consequently the final organic solvent percentage in the eluted product, was of key importance for the destabilization process, both in the degree of packing of the nanoparticles in the regular round structure and the final size of the clusters. In all the destabilization experiments presented, both for spherical and elongated structures, another polysaccharide, Cs, was used to "seal" and further condense the obtained structures. Actually, samples simply destabilized with ACN showed a reversible behavior after a single magnetic sorting cycle. The addition of the second, oppositely charged, polysaccharide, which strongly interacts with the external dextran layer, guarantees superior system stability.

The effects of the introduction of static magnetic fields in the destabilization process have been evaluated for the generation of structures with a high aspect ratio. In this case, the use of superparamagnetic nanoparticles allowed, using a dilute suspension of nanocrystals, not to generate any type of aggregates until the percentage threshold of ACN is reached: the formation of the elongated structure occurs very quickly, thus the strength and geometry of the applied magnetic field had a fundamental importance in the morphology of the superstructure. Weak fields (*C1* and *C2* configurations) produced non-regular (in shape and/or size) or highly aggregated superstructures. The best results were obtained with a magnetic gradient of 30.76 T m^{-1} (*P* configuration). As for spherical SMC, also for the EMS the addition of chitosan was fundamental to stop the destabilization step and to fix the final structure. The concentration and the timing of polysaccharide addition resulted in a drastic change in the elongated structures morphology.

By exploiting the high biocompatibility of all the components used, these destabilization and/or stabilization approaches through the use of paired polysaccharides can be used for multiple applications in the medical field.

5. Experimental Section

Materials: All solvents were purchased from Sigma-Aldrich and used without further modifications. $FeCl_2$ tetrahydrate, $FeCl_3$ hexahydrate and dextran (70KDa) were purchased from Carl Roth. Chitosan (low molecular weight, DA ≥ 75 %) was purchased from Sigma-Aldrich. The concentrator tubes used correspond to the Macrosep model of Pall Life Science. For the use of Cs, a stock solution was prepared at a concentration of 4% in water, with the addition of 1% v/v of pure acetic acid on the final volume.

Synthesis of magnetic nanoparticles (D-MNP): A single reaction step was followed for the synthesis of D-MNP: the aqueous solution containing iron salts and dextran is precipitated, by a progressive addition, drop by drop, of ammonium hydroxide at room temperature [19]. In detail, 318 mg of $FeCl_2$, 751 mg of $FeCl_3$ and 1.254 g of dextran were dissolved in 10 mL of ultrapure water and mixed with a stir bar. 10 mL of a solution of NH_4OH (7.5% in water) were added dropwise with a flow rate of 5 mL min^{-1}. The resulting black dispersion was still stirred for 5 minutes at RT; then the temperature was raised to 70 °C and kept for 30 minutes. The resulting brown dispersion was purified from aggregates by a fast centrifugation step (discarding the pellet). The excess of ammonium hydroxide was removed by five washing steps on concentrator tubes. The recovered magnetic nanoparticles were dispersed in a final volume of 8 mL of ultrapure water, obtaining an iron concentration of 0.3 M, and stored at room temperature (Figure A1).

Preparation of spherical nanoclusters - destabilization in tube: 20 μL of above-described D-MNP were diluted with 500 μL of ultrapure water and 500 μL of organic solvent were added dropwise (in 2 minutes) to induce the aggregation of the nanoparticles. Soon after, a solution of 2% Cs (in mixture water/solvent, at the same ratio used in the destabilization step) was added to the preformed clusters and mixed on an orbital shaker for 2 hours. Therefore, the Cs-stabilized magnetic nanoclusters were collected by magnetic separation and redispersed in water for characterization.

Preparation of spherical nanoclusters - destabilization in microfluidic device: a PDMS/glass microfluidic device was manufactured by using the standard soft lithographic technique for the exploitation of this approach [32]. The 2D cross-shaped flow focusing device was designed with a main channel/inlet, two orthogonal side branches/inlets and a single outlet. The microchannels had a height of 200 μm, whereas the width varied from 1200 μm in the inlet / outlet regions to 200 μm in the segments relating to the cross junction (see Figure A2). As described in ref. [32], oxygen-plasma surface activation was exploited during the fabrication assembly of the microfluidic device for PDMS-glass bonding. By this treatment, polar functional groups are introduced on PDMS and glass surfaces, changing their superficial assets. However, O_2-plasma treated surfaces undergo hydrophobic recovery after few hours. Plasma exposure resulted, as expected, in contact angle decrease: it changed from (107.19 ± 2.6)° to (52.02 ± 3.8)° and from (22.53 ± 4)° to 0°, for PDMS and glass respectively. After 24 h from plasma bonding, contact angle values were (72.89 ± 5)° and (12.49 ± 4.3)° for PDMS and glass respectively.

The solutions were continuously injected into the device by syringe pumps, with tailored injection scheme, flow rates values and concentrations described below in the Results section 2.2 and Table A3. The cross flow-focusing of ACN / water – water colloidal suspension phases was established on 24 h aged devices, in order to allow the same hydrophobicity degree to act on interfacial surface tension between the fluids and the channel walls.

Preparation of elongated nanoclusters: Destabilization in the presence of magnets was carried out using the following volumes: 5 μL of D-MNP 0.3 M, 645 μL ultrapure water, 350/400/450 μL of ACN. In the different tests carried out with magnets (*C1, C2, P*), the parameters relating to the different concentrations have been changed, for more details see Table A4-A6 in Appendix A. The addition of ACN was carried out manually, drop by drop, respecting the times shown in tables, keeping the system under constant stirring in a variable range from 200 rpm (orbital shaker Stuart SSL1) to 1400 rpm (orbital shaker Heidolph Multireax). As described for the spherical clusters, a solution of chitosan was added after the destabilization (after a time lapse described in Result section), at a concentration between 0.5 and 2%, dissolved in ACN 38% / water 62%. The Cs-coated nanostructures were left to shake overnight in the absence of permanent magnet, on an orbital shaker at 1400 rpm. The purification of obtained nanostructures was carried out by magnetic separation, redispersing the collected part in ultrapure water.

Magnetic gradient analysis: The nominal value and the magnetic gradient of each magnet were measured, by a bench gaussmeter equipped with a flexible probe, at a distance of 7.5 mm, in order to reproduce the configuration applied in EMS preparation. The results obtained show that for the single *C1* magnet the magnetic gradient is equal to 14.1 T m^{-1}, while for the tests carried out with *C1* magnets combined in multiple units, a value of 21.3 T m^{-1} is obtained for the four *C1* magnets and 16.7 T m^{-1} for the two coupled *C1* magnets. Regarding the *C2* magnets, the results show an increasing gradient from one to three units, respectively for single magnet the measured gradient is 12.4 T m^{-1}, with 2 units the measured value is 20.7 T m^{-1}; whereas with 3 units the measured value is 25.5 T m^{-1}; interestingly, coupling 4 *C2* units the magnetic gradient is equal to 25.8 T m^{-1}, a value that becomes saturated with the addition of further units, which shows that the power of the field does not increase further with the addition of new permanent magnets. The magnet *P* was instead used always as a single magnet (both for larger size and safety), with a measured magnetic gradient value of 30.76 T m^{-1}.

Instrumentation: The morphology and size distribution of individual magnetic nanoparticles and superstructures were analyzed by a transmission electron microscopy (TEM) system (JEOL JEM 1011) operating at 100 kV, equipped with a TEM CCD camera Mega View III. Samples were drop-casted on 300 mesh carbon-coated copper grid for the analysis. Hydrodynamic diameter and surface charge of clusters were evaluated by dynamic light scattering (Malvern Zetasizer Nano ZSP).

6. Appendix A

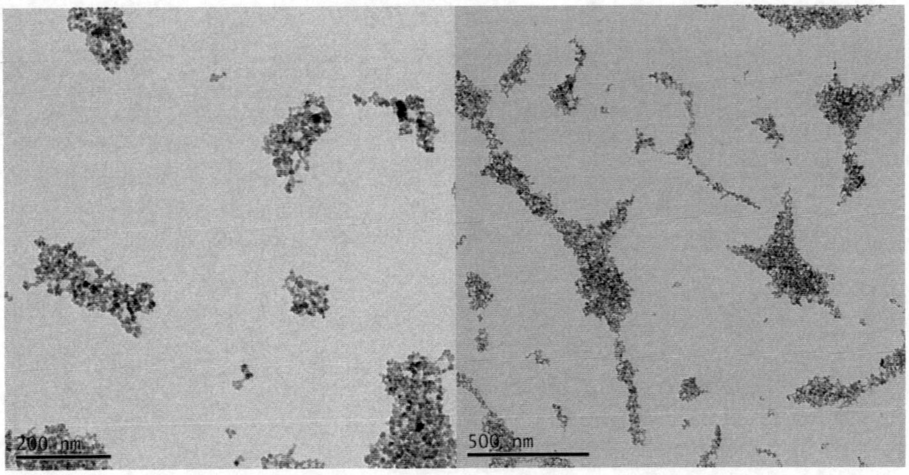

Figure A1. TEM images of dextran-coated magnetic nanoparticles.

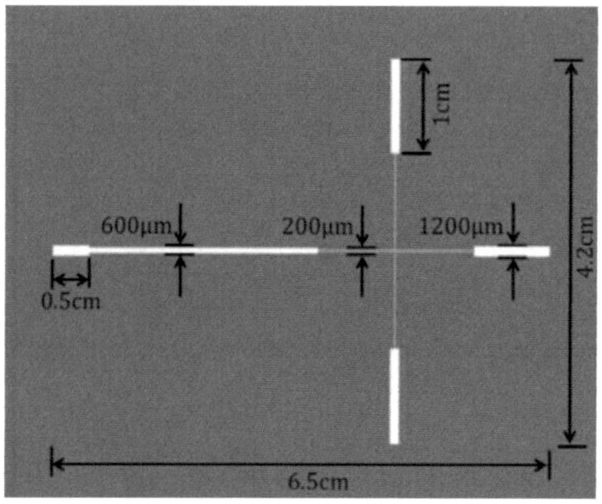

Figure A2. Cross-shaped channel pattern of the microfluidic device. Channel geometry in terms of length and width are here indicated. Height of the manufactured channels, 200 µm.

Table A1. Destabilization experiments with different organic solvents. A preliminary experiment as performed to macroscopical evaluate the aggregation effect of different organic solvent in the tube destabilization of a dextran-coated magnetic nanoparticles suspension. The partial solubility of dextran in the different solvents (all miscible with water) drove the aggregation grade obtained. In details, acetone and ethanol caused a fast and severe aggregation, due to the dextran insolubility in monohydric alcohols and ketones. On the opposite, a solvent as dimethylformamide did not caused any aggregation because the polysaccharide is quite soluble in formamide. A polar aprotic solvent as acetonitrile was selected as best choice to control the nanoparticle destabilization in subsequent experiments.

H_2O (µL)	Magnetic NPs (0.3 M, µL)	Solvent (µL)	Results
500	20	500 - Ethanol (P. index 5.2)	relevant aggregation
500	20	500 - Acetonitrile (P. index 5.2)	mild and stable aggregation
500	20	500 - Acetone (P index 5.1)	intense aggregation
500	20	500 - Dimethylformamide (P. index 6.4)	no effect *

* Dextran is soluble in dimethylformamide, no destabilization occurred

Table A2. Summary of in tube destabilization tests. Best conditions highlighted in green.

test	acetonitrile (μL)	H_2O (μL)	Magnetic NPs (0.3M, μL)	Cs%/ACN%
1	400 (40% $_{final}$)	600	20	1% / 38%
2	340 (35% $_{final}$)	600	20	1% / 38%
3	300 (30% $_{final}$)	600	20	1% / 38%
4	400 (38% $_{final}$)	645	5	1% / 38%
5	400 (38% $_{final}$)	645	10	1% / 38%

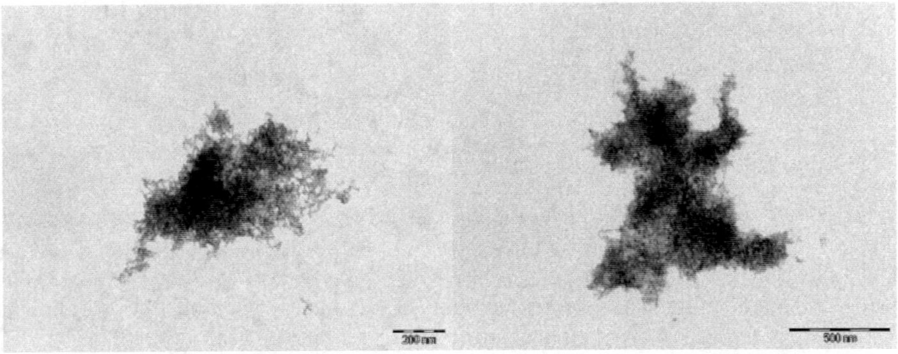

Figure A3. TEM images of sample 1 (in tube destabilization, Table A2)

Table A3. Summary of the injection optimization experiments for device destabilization and successive coating. The magnetic nanoparticles were injected by the two orthogonal side branches; thus, the overall addition must be considered as doubled (2x) referring to the single flow rate. Best conditions highlighted in green.

test	Magnetic NPs 2.5mM (ml min^{-1}) (X2)	Acetonitrile (ml min^{-1})	Side/Core flow rate ratio	Stabilization step (Cs coating out of the device)
1	0.3	0.4	1.5	Chitosan (0.5%) / Acetonitrile 38%
2	0.275	0.45	1.2	Chitosan (0.5%) / Acetonitrile 38%
3	0.325	0.35	1.86	Chitosan (0.5%) / Acetonitrile 38%
4	0.35	0.3	2.33	Chitosan (0.5%) / Acetonitrile 38%
5	0.25	0.5	1	Chitosan (0.5%) / Acetonitrile 38%
6	0.275	0.45	1.22	none
7	0.275	0.45	1.22	Chitosan (1%) / Acetonitrile 45%
8	0.275	0.45	1.22	Chitosan (1%) / Acetonitrile 50%

IOP Conf. Series: Materials Science and Engineering 1265 (2022) 012006 doi:10.1088/1757-899X/1265/1/012006

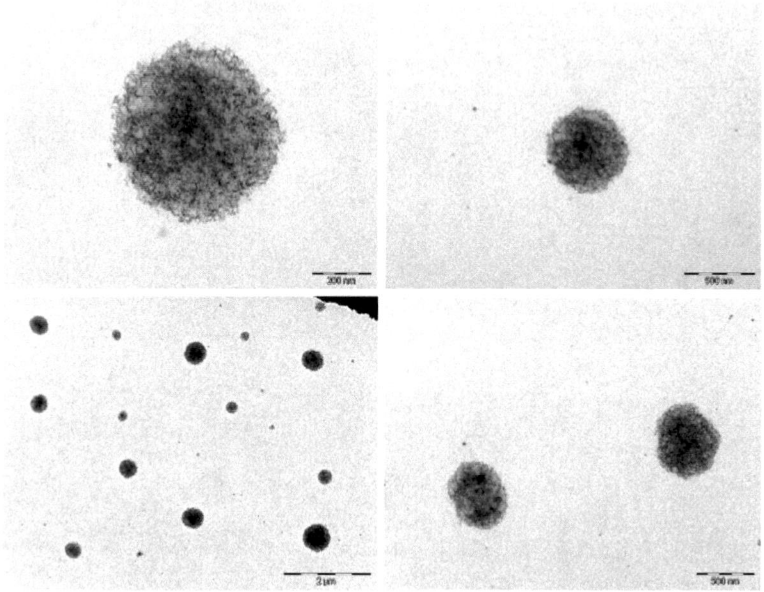

Figure A4 TEM images of SMC obtained by injection of magnetic nanoparticles at 0.275 mL min^{-1} and acetonitrile at 0.45 mL min^{-1}. Since no stabilization step was applied to the obtained SMC, the grid was dried in a saturated acetonitrile chamber, in order to preserve the 3D-structure of the cluster. Otherwise, the fast evaporation of the solvent could induce the disassembly of the nanoparticles. The SMC were drop-casted without any magnetic purification after the destabilization.

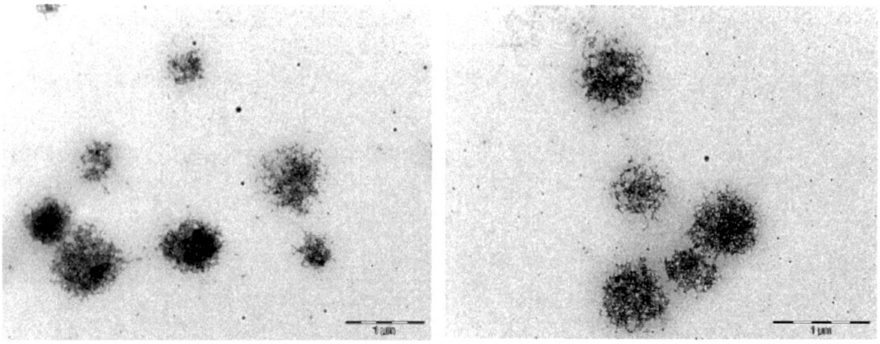

Figure A5. TEM characterization of SMC obtained after destabilization of magnetic nanoparticles (injection at 0.275 mL min^{-1}) with acetonitrile (injection at 0.45 mL min^{-1}) and subsequent stabilization in Cs (1%) / ACN 45%.

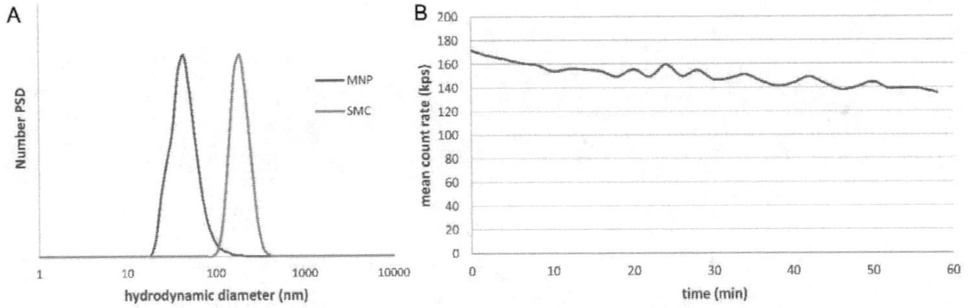

Figure A6. DLS analysis of SMC. Panel A) hydrodynamic diameter comparison between magnetic nanoparticles (blue curve) and SMC (orange curve). Panel B) mean count rate variation analysis of SMC dispersed in cell culture medium, in a 60 minutes time lapse (single read every 2 minutes).

Table A4. Summary of EMS preparation tests with magnet *C1*. Best conditions highlighted in green.

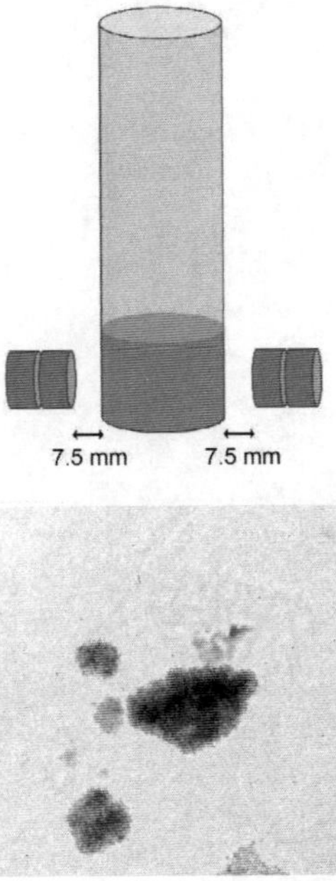

test	Acetonitrile (µL)	Cs% / waiting time	Vortexing speed (rpm)
A	400 µL	2% / 1'	700
B	400 µL	1% / 1'	400
C	400 µL	0.5%/ 1'	1000
D	400 µL	0.5%/ 1'	1400
E	350 µL	1% / 1'	1400

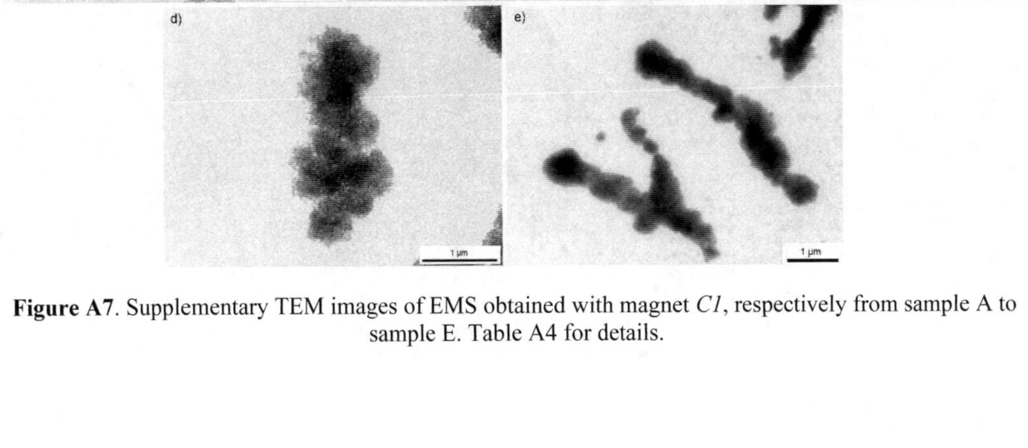

Figure A7. Supplementary TEM images of EMS obtained with magnet *C1*, respectively from sample A to sample E. Table A4 for details.

Table A5. Summary of EMS preparation tests with magnet *C2*. Best conditions highlighted in green.

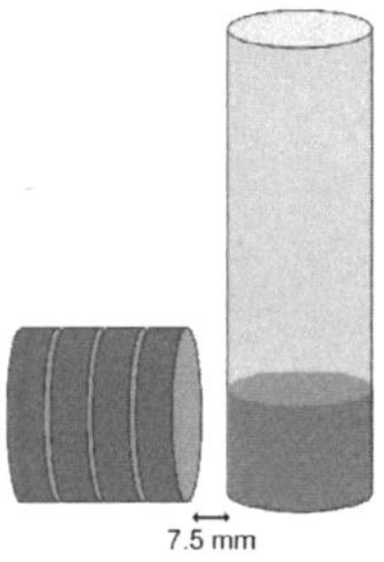

test	Number of magnets	acetonitrile	waiting time / 2% Cs volume
A	4	400 μL	5' / 1 ml
B	5	400 μL	5' / 1 ml
C	3	400 μL	5' / 1 ml
D	4	350 μL	5' / 1 ml
E	4	400 μL	5' / 2 ml

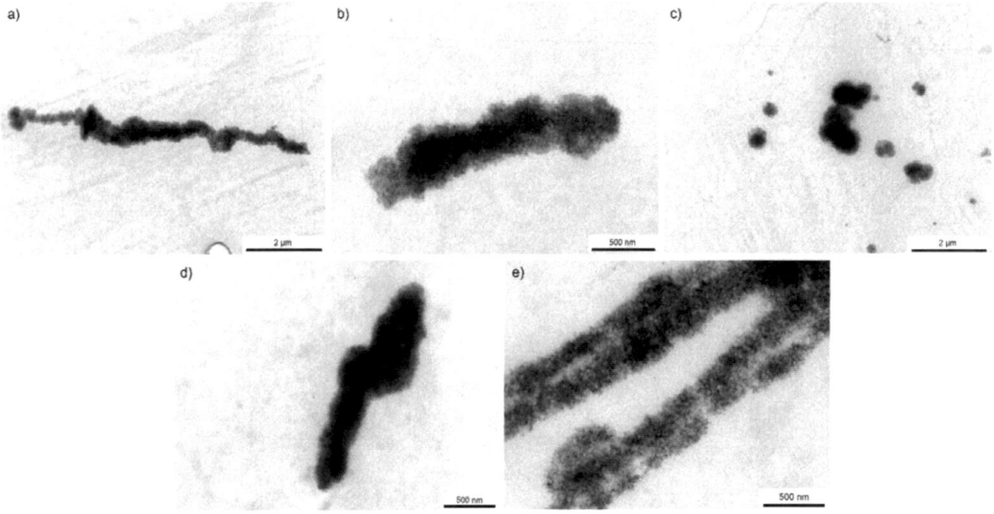

Figure A8. Supplementary TEM images of EMS obtained with magnet *C2*, respectively from sample A to sample E. Table A5 for details.

IOP Conf. Series: Materials Science and Engineering 1265 (2022) 012006 doi:10.1088/1757-899X/1265/1/012006

Table A6. Summary of EMS preparation tests with magnet *P* in horizontal configuration. Best conditions highlighted in green.

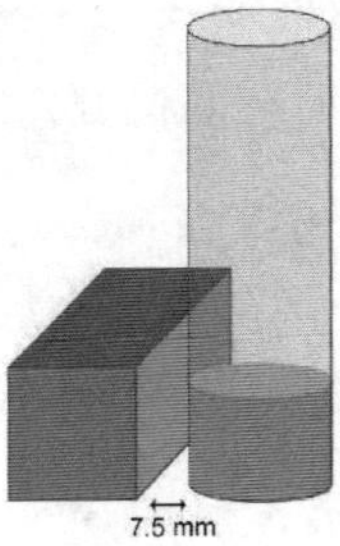

test	acetonitrile	Cs % / waiting time
A	350 μL	2% / 1’
B	350 μL	2% / 5’
C	350 μL	2% / 15’
D	400 μL	2% / 5’

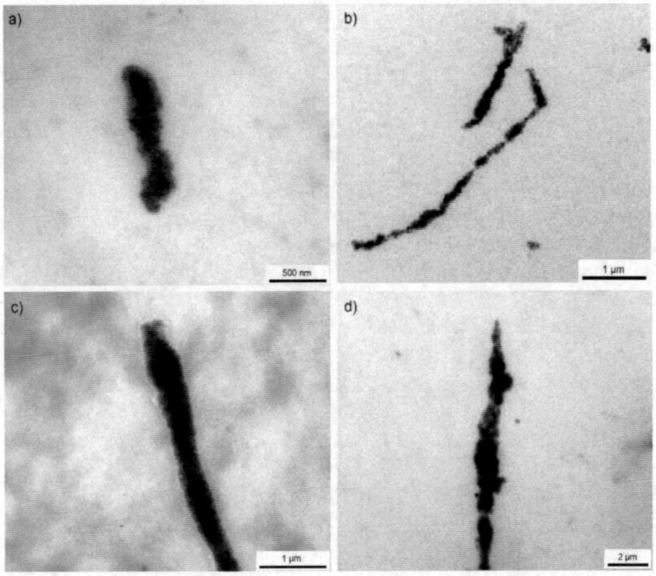

Figure A9. Supplementary TEM images of EMS obtained with magnet *P* in horizontal configuration, respectively from sample A to sample D. Table A7 for details.

References

1. Avasthi A, Caro C, Pozo-Torres E, Leal MP and García-Martín ML 2020 *Topics in Current Chemistry* **378** 40.
2. Ibrahim Y 2016 *Nanotechnology Reviews* **5** 331-40.
3. Wu K, Su D, Liu J, Saha R and Wang J-P 2019 *Nanotechnology* **30** 502003.
4. Athanasia K and Alexandros L 2015 *Nanotechnology Reviews* **4** 595-624.
5. Mandriota G, Di Corato R, Benedetti M, De Castro F, Fanizzi FP and Rinaldi R 2019 *Acs Applied Materials & Interfaces* **11** 1864-75.
6. Di Corato R, Bealle G, Kolosnjaj-Tabi J, Espinosa A, Clement O, Silva AKA, Menager C and Wilhelm C 2015 *Acs Nano* **9** 2904-16.

7. Di Corato R, Bigall NC, Ragusa A, Dorfs D, Genovese A, Marotta R, Manna L and Pellegrino T 2011 *Acs Nano* **5** 1109-21.
8. Paquet C, de Haan HW, Leek DM, Lin H-Y, Xiang B, Tian G, Kell A and Simard B 2011 *ACS Nano* **5** 3104-12.
9. Albarqi HA, Wong LH, Schumann C, Sabei FY, Korzun T, Li X, Hansen MN, Dhagat P, Moses AS, Taratula O and Taratula O 2019 *ACS Nano* **13** 6383-95.
10. Berret JF 2016 *Nature Communications* **7** 10134.
11. Mazuel F, Mathieu S, Di Corato R, Bacri JC, Meylheuc T, Pellegrino T, Reffay M and Wilhelm C 2017 *Small* **13** 1701274.
12. Rosen JE, Chan L, Shieh D-B and Gu FX 2012 *Nanomedicine: Nanotechnology, Biology and Medicine* **8** 275-90.
13. Wáng YXJ and Idée J-M 2017 *Quantitative Imaging in Medicine and Surgery* **7** 88-122.
14. Di Corato R, Aloisi A, Rella S, Greneche JM, Pugliese G, Pellegrino T, Malitesta C and Rinaldi R 2018 *ACS applied materials & interfaces* **10** 20271-80.
15. Basiruddin SK, Ranjan Maity A and Jana NR 2012 *RSC Advances* **2** 11915-21.
16. Park YI, Piao Y, Lee N, Yoo B, Kim BH, Choi SH and Hyeon T 2011 *Journal of Materials Chemistry* **21** 11472.
17. Kim M-H, Kim B, Lim E-K, Choi Y, Choi J, Kim E, Jang E, Park HS, Suh J-S, Huh Y-M and Haam S 2014 *Macromolecular Bioscience* **14** 943-52.
18. Park JC, Park TY, Kim DH, Cha HJ and Seo JH 2020 *Chemical Engineering Journal* **379** 122406.
19. Lüdtke-Buzug K, Biederer S, Sattel T, Knopp T and Buzug TM 2009 *4th European Conference of the International Federation for Medical and Biological Engineering* **22** 2343-6.
20. Quiñones JP, Peniche H and Peniche C 2018 *Polymers* **10** 235.
21. Schatz C, Lucas J-M, Viton C, Domard A, Pichot C and Delair T 2004 *Langmuir* **20** 7766-78.
22. Drogoz A, David L, Rochas C, Domard A and Delair T 2007 *Langmuir* **23** 10950-8.
23. Schatz C, Domard A, Viton C, Pichot C and Delair T 2004 *Biomacromolecules* **5** 1882-92.
24. Roman DL, Roman M, Som C, Schmutz M, Hernandez E, Wick P, Casalini T, Perale G, Ostafe V and Isvoran A 2019 *Frontiers in Bioengineering and Biotechnology* **7** 214.
25. Leyre P-Á, José Manuel L and Arturo Á-B 2016 *Current Pharmaceutical Design* **22** 3380-98.
26. Zhang Y, Sun T and Jiang C 2018 *Acta Pharmaceutica Sinica B* **8** 34-50.
27. Toma CC, Aloisi A, Bordoni V, Di Corato R, Rauner M, Cuniberti G, Delogu LG and Rinaldi R 2018 *Biomacromolecules* **19** 3560-71.
28. Martel JM and Toner M 2014 *Annual Review of Biomedical Engineering* **16** 371-96.
29. Kilinc D, Dennis CL and Lee GU 2016 *Adv Mater* **28** 5672-80.
30. Martínez-Banderas AI, Aires A, Teran FJ, Perez JE, Cadenas JF, Alsharif N, Ravasi T, Cortajarena AL and Kosel J 2016 *Scientific Reports* **6** 35786.
31. Navya PN, Kaphle A, Srinivas SP, Bhargava SK, Rotello VM and Daima HK 2019 *Nano Convergence* **6** 23.
32. Aloisi A, Toma CC, Corato RD and Rinaldi R 2017 *Proc. SPIE 10364, Organic Sensors and Bioelectronics X* **10364** 103640O.

Acknowledgements
This research was funded by Italian PON project AMIDERHA (PON02_00576_3329762) and by BIO-D project (ARS01_00876) co-financed by European Union—ERDF – ESF – PON Research and Innovation 2014–2020.

The authors thank the NaBiDit Laboratory ("Nano-Biotechnological methods for innovative Diagnostics and Therapy", Università del Salento) for the access to facility resources.

Safe by Design of nanomaterials in industrial production processes: nanomaterials in the cosmetic field

S Vertuani[1,2], E Cesa[1], M Battistin[3], P Ziosi[2] and S Manfredini[1,2]

[1] University of Ferrara, Department of Life Science and Biotechnology, Via L. Borsari 46 44121-Ferrara, Italy
[2] Ambrosialab Srl, accredited Spin-off company of the University of Ferrara, Italy
[3] Kalis Srl, Onigo di Pederobba (TV), Italy

Corresponding author: smanfred@unife.it

Abstract. Cosmetic pleasantness of products containing inorganic UV filters depends on an unwanted whitening effect which is inversely correlated to their size. To overcome this effect, diameter has been reduced to nanosized but thus increasing the surface area and biological reactivity and carrying questions about the safety of humans and environment. Following a "safe and sustainability by design" approach, we designed a new class of optimized sunscreen UV filters by chemical functionalization of ZnO and TiO_2 with Oxisol. If compared with the simple physical mixture, the new coated filters show different properties and benefits: a higher SPF (ISO 24443:2012), a better cytotoxic profile (MTT and NRU assay), a radical scavenging action (PCL assay) and an improved safety profile with a strong reduction of photocatalytic activity (Acid blue 9 test). In conclusion, our Safe by Design (acronym SbD) approach presents a new generation of UV filters bonded with booster molecules (by means of synergistic antioxidant effects) as the best compromise in the conscientious UV protection.

Keywords: TiO_2 nano, ZnO nano, inorganic UV filters, Safe by design (SbD), nanoparticles, booster

1. Introduction

In recent decades, the precaution against the unwanted skin effects from the sun radiation has played a key role in terms of human skin health but additionally involved into a public policy concern.

The ultraviolet (UV) rays are the most harmful radiation to the skin and an effective photoprotection against sunburning is essential to avoid skin disorders primarily in terms of photo-carcinogenesis, but also photo-immunosuppression and photo-aging. In fact, each of the UV components explain damage to cells, tissue, and molecules.

Fortunately, the UVC (100-280 nm) component with the highest energy is absorbed by stratospheric ozone and consequently the sunlight reaching the Earth's surface consists predominantly of 90-95% UVA (320-400 nm) and 5-10% UVB (290-320 nm) radiation, both enable to penetrate the skin and cause biological and metabolic reactions [1].

Nowadays, the climate conditions (i.e., the low presence of atmospheric particulate and aerosol pollution to scatter and reflex UV) and the depletion of the ozone layer by the chlorofluorocarbons, halons, and nitric oxides are worried about the possible result in an increased irradiance level of UVB and UVC at the Earth's level with the consequence of a higher incidence of harmful effects of all life forms.

Several complementary approaches should be internalized to prevent health issues such as protection through clothing and the use of topical sunscreens filters.
In particular the arguments of UV filters continue to be frequently discussed, being highly regulated and limited by different national legislations. For example, in Europe an UV filter as an ingredient of personal care products requires the positive assessment by the independent EU advisory Scientific Committee for Consumer Safety (SCCS) before commercialization in the EU. After that, there is an approval by the EU Commission with the inclusion into the Annex VI of the Cosmetic Regulation n° 1223/2009, allowing the use in the European Union (EU) [2].
Different is the situation in the United State, where sunscreens are framed in a different way. They are classified as over-the-counter (OTC) and they follow the strict regulation of the US Food and Drug Administration (FDA), according to the pharma requirements.
According to their physicochemical properties, the UV sun blockers can be classified as organic or inorganic and they work by adsorption and/or reflection of UV radiation.
They should provide protection against the adverse effects of both UVB fully adsorbed by the stratum corneum and the top layers and UVA radiation whereas up to 50% penetrates Caucasian dermis.

2. Disputes regarding UV filters in sunscreens

2.1. The Burning issues of organic UV filters

The class of organic UV filters have become controversial due to their potential impact on the environment with the coral bleaching and their prevalence in the water supply and in aquatic animals. Following a public debate, on May 2018, firstly Hawaiian state legislature and subsequently other fascinated destination such as Palau and the U.S. Virgin Islands passed a bill banning the sale and distribution of sunscreens containing organic filters benzophenone-3 (BP3, also known as oxybenzone) and ethylhexyl methoxycinnamate (EHMC or octinoxate) [3], while promoting the use of inorganic filters. Subsequently, the bill became law in January 2021 [4].
These bans were based on preliminary scientific studies but without standardized testing schemes for scleractinian corals and the lack of hard scientific evidence have led to the emergence of a new sunscreen market claim for product "coral reef safe" or "reef friendly". A clear definition and regulation of this topic are still lacking [5].
Furthermore, organic sun blockers have been reported to have negative hormonal effects in animal models and their potential human health risks as endocrine disruptors. Often there is a generalization, because not all organic sunscreens are harmful, but the market wants inorganic sunscreens that are becoming ever more promising candidates for photoprotection.

2.2. The growing role of the Inorganic UV filters as safer alternatives

Inorganic sunscreens are basically two, titanium dioxide (TiO_2) and zinc oxide (ZnO), boasting a lack of percutaneous adsorption and absence of skin irritation, inertness of the ingredients, a broad spectrum protection, high performance, and more eco-friendly profile than organic sunblokers [6]. Their UV attenuation results from different mechanisms including reflection and scattering of UV radiation and visible light (clarifying the white effect of these sunscreens formulation) and UV adsorption.
TiO_2 is more effective in UVB protection and ZnO in UVA range but with a broad absorption in all UV ranges. For this reason, usually a combination of mineral particles assures broad-band UV protection [7].
To solve the cosmetic drawback of their natural opaqueness, microsized inorganic UV filters have been increasingly replaced by nanoparticles (NPs) with core dimension minors of 100nm, reaching the optimal light scattering size approximately half the wavelength. This aspect leads to a more transparent film on the skin after application which enhances consumer's acceptance and a good spread ability to be used.
Besides avoiding the white effect, the dimension aspects also affect the photoprotection performance of

ZnO and TiO_2 NPs performing better than the non-nano (more surface area) that allows use of less concentration. However, the popularity and the widespread use of these ZnO/ TiO_2 based sunscreens have increased concerns regarding their safety, related mainly to the nano toxicity issue.

2.2.1. Challenge of Inorganic NPs in sunscreens. Since the surface area to volume ratio of particles increases as the particle diameter decreases, NPs are under scrutiny for probable (bio)reactive issues compared to normal bulk materials.

The most potential risk about nanomaterials is about the exposure via inhalation, prompting recommendations against spray sunscreen products with NPs.

In particular, the issue associated with inorganic nano sunscreen is related to phototoxicity for TiO_2 and cytotoxicity for ZnO, concerning the higher surface reactivity and bioavailability.

Even though the presence of NPs in sunscreen is unlikely to cause more trouble than sun exposure itself, there is a tricky point in the European cosmetic regulatory framework.

In article 2(1) of the EU Cosmetics Regulation, a nanomaterial is defined as "an insoluble or biopersistent and intentionally manufactured material with one or more external dimensions, or an internal structure, on the scale from 1 to 100 nm " [2]. In this way, everything should follow this characterization.

Subsequently the EU Commission adopted a recommendation on the definition of a nanomaterial, describing as "a natural, incidental, or manufactured material containing particles, either in the free state or as an aggregate or agglomerate and in which, for 50% or more of the particles in the number size distribution, one or more external dimensions is in the size range 1 nm-100 nm" [8]. This latter different definition was applied in the REACH regulation thus creating discrepancies: a material could be classified as a nanomaterial under the REACH regulation, but not under the EU Cosmetics Regulation. In June 2022, to align legislation across all sectors, the European Commission has clarified the definition of nanomaterials in a new recommendation. The new recommendation, a result of the Chemicals for Sustainability Strategy, supports a coherent European regulatory framework for nanomaterials, helping to align legislation across all sectors. This definition, replacing the previous one from 2011, should be used in EU and national legislation, policy or research programs. The changes were developed following a comprehensive review and should allow for easier and more efficient implementation, but will not significantly affect the scope of identified nanomaterials [9].

By the new definition, *"Nanomaterial' means a natural, incidental or manufactured material consisting of solid particles that are present, either on their own or as identifiable constituent particles in aggregates or agglomerates, and where 50 % or more of these particles in the number-based size distribution fulfill at least one of the following conditions: (a) one or more external dimensions of the particle are in the size range 1 nm to 100 nm; (b) the particle has an elongated shape, such as a rod, fiber or tube, where two external dimensions are smaller than 1 nm and the other dimension is larger than 100 nm; (c) the particle has a plate-like shape, where one external dimension is smaller than 1 nm and the other dimensions are larger than 100 nm. In the determination of the particle number-based size distribution, particles with at least two orthogonal external dimensions larger than 100 μm need not be considered. However, a material with a specific surface area by volume of < 6 m2 /cm3 shall not be considered a nanomaterial".*

3. Experimental details

3.1. Our SbD approach: Case study

To overcome this debate, our approach has investigated the NPs sunscreen safety from a physicochemical point of view, able to influence both sunscreen efficacy and safety.

For example, a higher interfacial energy causes phenomena of aggregations, partly determined by

surrounding conditions like pH and ion strength, with consequences on NPs size and the band gap [10]. A smaller band gap energy reduces the amounts of energy required to overcome the gap, corresponding to higher wavelengths, with a modulation of the NP UV attenuation properties.
Our coating designs started from small molecules provided with intrinsic antiradical activity to generate sustainable sunscreens. This approach is the base of SbD projects, aimed at the design, since the very beginning, of an intrinsically safe material safe and sustainable for the environment. Current approaches usually work on the other way around firstly obtaining the compounds and only later considering their potential drawbacks. As an example, to prevent the potential risks of NPs based sunscreens, the type of TiO_2 and ZnO NP coating and the kind of formulation strategy should be considered at first. Formulation strategy should be considered to prevent NP aggregation, enhance sunscreen photostability, and increase sunscreen efficacy but not the cutaneous permeability, to avoid possible NPs absorption. This in parallel to concerns related to the important role of NPs coating in controlling behavior in suspension and surface reactivity.

4. Materials and methods

4.1. Functionalization of TiO_2 and ZnO

A solution, approved by EU cosmetic regulation (Annex VI) has already been introduced in the market, as an alternative made by coating with internal materials (e.g., with silica and others) that improve the performance of the UV filters and increase their dimension, but still requires the INCI (nano) definition on the label by the European cosmetic regulation, since the core is nano-sized.
The photoactivity of the particles can be influenced by their surface characteristics. With this purpose, coatings are applied to prevent adverse effects mediated by photocatalytic redox reactions at the NPs' surface, including the degradation of organic formulation components by produced ROS. As a result of coating, the NPs interaction is interrupted (with formulation, environment, and skin).
Commonly used sunscreen NP coating that does not affect UV attenuation comprise silica, aluminum oxide, aluminum hydroxide, methicone, and polymethylacrilyc acid.
These materials function by capturing reactive radicals or inhibiting their formation by preventing contact between TiO_2 surface, oxygen, and water. The coating, however, does not guarantee the absence of photocatalytic activity [11].
Recently, Ambrosialab, an accredited spinoff company of the University of Ferrara, has developed a polyphenol, Oxisol [12,13], endowed by potent free radical scavenging activity and characterized by a catechol and a carboxylic acid functionality (Figure 1).

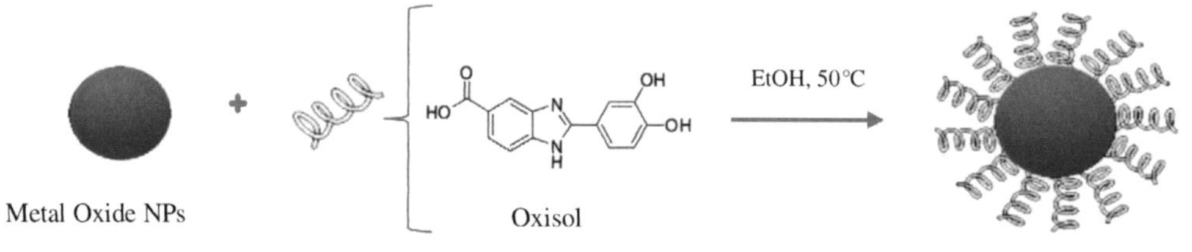

Figure 1. Schematic representation of functionalization of metal oxide NPs and Oxisol.

This behaves as an active coating, able to develop important activities: catechols are very potent antioxidant molecules, very diffuse in nature. Using this molecule as a coating agent, we obtained a suppression of the photoreactivity, without loss of the sunscreen activity in the respect of the uncoated molecule but, surprisingly, with a synergistic effect. At present the exact mechanisms at the base of this observed behavior are still under investigation.

4.2. *Safety test of inorganic NPs*

The problem of toxicity is related to humans and the environment. For this purpose, our approach was designed as a SbD strategy in an active collaboration with ITENE (Instituto Tecnológico del Embalaje, Transporte y Logística, Valencia, Espania) and Sabanci University Nanotechnology Research and Application Center (SUNUM), Tuzla, Istanbul, Turkey.

4.2.1. *Cytotoxicity*. The cytotoxicity test was measured by neutral red assay (NRU) using 3T3 cells, performing in DMEM (Dulbecco's modified Eagle medium) supplemented with 10% fetal calf serum (FCS), penicillin and streptomycin (both 100 U/mL) and glutamine (2mM). Cells were treated for 48 h with different concentrations (1μg/mL, 10μg/mL, and 100μg/mL) and then washed and added 250 μL of NRU solution (25μg/mL). After 2h, cells were washed again and 150 μL of NRU-desorb solution (49% water, 50% EtOH and 1% glacial acetic acid) was added. Finally, the absorbance of the solution after 40 min was measured at 540 nm. The value obtained is proportional to the number of live cells and it can be converted into % of growth inhibition [12,13].

4.2.2. *Photoreactivity for TiO_2: Blue Acid 9 assay.* One of the first analyses conducted regarded the photocatalytic activity (first issue for the toxicity generated from the TiO_2 especially in the nanoform). Titanium dioxide has three different crystalline structures: anatase, rutile and brookite. Rutile is the most common and stable form, while anatase is the most photoreactive. TiO_2 is a semiconductor material with an electronic structure characterized by bands and orbital separated by energy band gaps for which there are no orbitals. Upon light adsorption of minimally the band gap energy between the Valence band (VB) (for TiO_2 formed by the O-*2p* states) and the conduction band (CB) (for TiO_2 formed by the Ti-*3d* state), a VB electron (e^-) is excited to the CB leaving a hole (h^+) of a highly localized electron vacancy in the VB of the semiconductor particle.[13] The redox reactions on the crystal's surface of these particles are due to the migration of both e^- and h^+. Effects are predominantly mediated by formation of superoxide anion radicals ($O_2{-}\bullet$) and hydroxyl radicals (•OH) in the presence of light. However, the production of singlet oxygen ($^1O_2$) and H_2O_2 may play a role as well. Anatase generally displays the highest photoactivity.
The value of Band gap energy is equal to 3.03 eV for rutile bulk and 3.2 eV determined for anatase. To understand if the coating works to prevent the photocatalytic reactivity and consequently increase the stability, we have studied the photodegradation pattern against a dye (Blue Acid 9) as a mimic of a free radical generated from UV sun radiation that hit the inorganic filters [14].
10 mg of both pure metal oxide and uncoated metal oxide was added to 100mL of a 0.77 mM ethanolic solution of Blue Acid 9. Under magnetic stirring, one mixture was exposed to UVA radiation (8W lamp at 15cm distance) at room temperature and another one in dark condition to remove the contribution of dye adsorption. After 1h, each sample is left to rest for 3h, then centrifuged and filtered to be analyze with the UV-vis spectrophotometer (628nm).

4.2.3. *SPF and antioxidant activity* The *in vitro* method used in this work was recently proposed by us, adapting the ISO 24443:2012 standard for the determination of UVA protection to UVB [15]. In vitro SPF spectrophotometric evaluation was performed measuring transmittance spectrum by a SHIMADZU UV-2600 spectrophotometer provided of integrating sphere ISR 2600 60 mm and coupled with an SPF determination software (SPF calculator software version 2.1, Shimadzu, Milan, Italy) to obtain the values of SPF, UVA/UVB, UVAPF, and λ critical. We used an *in vitro* approach used consisted of applying a thin film of product on an artificial substrate that must be as similar as possible to human skin, PMMA plates (Schonberg GmbH, Munich, Germany) with a 5 μm reproducible roughness and an area of 25 cm^2. Via

spectrophotometric measures, the amount of UV radiation passing through the filmcan be evaluated. Photostability studies were carried out with a solar simulator device (Suntest CPS; Atlas, Linsengericht, Germany) equipped with a Xenon lamp, an optical filter to cut off wavelengths shorter than 290 nm, and an IR-block filter to avoid thermal effects, as prescribed by the ISO 24443:2012 procedure.
In order to evaluate the antioxidant activity of a sample against superoxide anion radicals, we performed the PCL. This test is based on Popov and Lewin's method [16] and it measures the superoxide anion radicals generated from luminol when exposed to UV light (Double Bore® phosphor lamp, output 351 nm, 3 mWatt/cm^2). Measurements were carried out with a Photochem® apparatus (Analytik Jena, Leipzig, Germany) and the antioxidant capacity was measured using Antioxidant Capacity of Liposoluble (ACL) substance kits [17] and work solution of Luminol reagent and Trolox (standard) were freshly prepared according to the ACL protocol. The kinetic light emission curve was monitored for 180s and expressed as micromoles of Trolox® per gram of dry matter. The areas under the curve were calculated using the PCLsoft control and analysis software, using a A calibration curve of scalarTrolox® concentrations. The different reduction in signal amplitude was used as a parameter for quantification and correlated to the reduction of the integral of PCL intensities caused by the different concentrations of Trolox®. The results of antioxidant activity were expressed in micromoles of Trolox®, which provide an antioxidant capacity equivalent to 1 g of the sample under examination.

5. Results and Discussion

5.1. Photocatalytic evaluation

In the comparison between the nano and the nano coated materials we have observed an impressive reduction of reactivity. There are no photocatalytic activities anymore in coated NPs.
This evaluation also reflected the higher photocatalytic activity of TiO_2 anatase compared with ZnO and emphasized the importance of the type of coating (Figure 2).

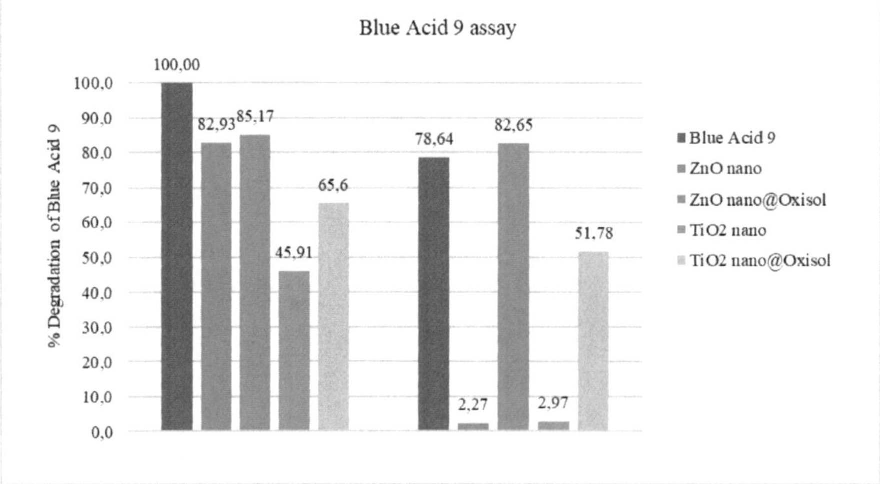

Figure 2. Photocatalytic activity of samples expressed as % degradation of Blue acid 9 both in condition of light and darkness.

The results observed were very interesting and we have different hypotheses on the mechanism: (a) a chemical hypothesis related to an injection of electrons on the metal oxide conduction band by the coating and this reduces the photochemical activity or (b) steric hindrance (Figure 3).

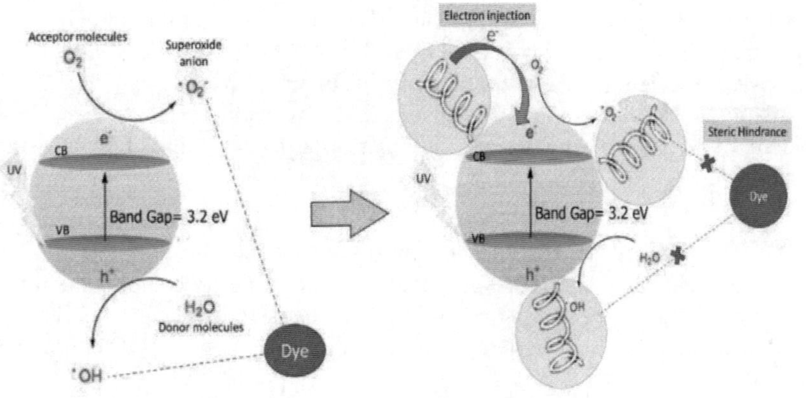

Figure 3. Hypothesis of photocatalytic quenching: (a) injection of electrons and (b) steric hindrance.

5.2. *Cytotoxicity on 3T3 cells treated for 48h*

Compared with TiO_2, ZnO appears to be less photoreactive but may display, unlike TiO_2, instability during reaction. In particular, ZnO has a high solubility in water and the dissolved Zn^{2+} species are more bioavailable than ZnO itself, increasing the potential risk notably in seawater ecosystems, this issue is not considered by most of the authors attributing more risks to ZnO than the real ones. Indeed, before the recent opinion by EFSA on genotoxicity of TiO_2 [13], TiO_2-based UV filters were generally viewed as safer due to their weaker solubility in aqueous media.

Following our approach, we have studied the cytotoxicity before the formulation study considering the cytotoxicity related to ZnO and that attributed to TiO_2 (Table 1).

Table 1 Results of cytotoxicity test as cell growth inhibition at different concentration of compounds (1,10,100 µg/ml).

	Concentration (µg/mL)	*% Inhibition*	*± sd*
Control	0	0	0
ZnO nano	1	0,15	6,56
	10	61,11	12,3
	100	87,89	0,81
TiO₂ nano	1	−2,73	1,78
	10	−7,50	1,04
	100	3,08	5,24
ZnO nano@Oxisol	1	−9,33	3,8
	10	4,2	4,56
	100	88,02	0,8
TiO₂ nano@Oxisol	1	−1,37	4,27
	10	0,18	4,8
	100	4,17	6,2

In these conditions TiO_2 was not toxic but ZnO showed some cytotoxic effects, even at the highest concentrations used. However, the latter toxicity is due to the release of Zn^{2+} and not to ZnO itself. No in vivo data have ever highlighted ZnO toxicity itself. In addition, ZnO nano is not "nano" for the EU cosmetic regulation because it is not bio-persistent and it is readily dissolved when in the body. Thus, we concluded that reduction of ZnO toxicity was mainly due to the ability of Oxisol to complex released Zn^{2+}.

5.3. SPF and antioxidant activity

The antioxidant capacity was reported as µmol TE/gram, in order to evaluate the influence of coating agent since the activity is related to the amount of free hydroxyl phenolic groups. The free radicals were visualized with a chemiluminescence detection reagent (luminol) that acts as photosensitizer as well as oxygen radical detection reagent.

As reported in Figure 4, the antioxidant activity of TiO_2 covered materials increased in an impressive way, this consequently allows us to consider them as multifunctional ingredients with a synergistic activity toward UV ray damages that can be counteracted by both reflection-absorption mechanism and also by ROS scavenging. On the other hand, for ZnO, functionalization negatively affects antioxidant capacity, leading many to argue that the formation of the complex inactivates Oxisol.

Surprisingly, the coating not only maintained but also increased the sunscreen properties of the starting ZnO and TiO_2 materials. In particular, the sun protection activity of TiO_2 nano@Oxisol almost doubled for SPF in comparison to mixture TiO_2 nano+Oxisol with an increase of 73%.

For the product ZnO nano@Oxisol, it was registered a consistent SPF value in comparison with the simple mixture: the covalent bonding increases the SPF value by 107% (from 2.01 to 4.17). In our hypothesis, the coordination of ZnO could lead to a new raw material with new optical properties, not comparable with the simple mixture. Finally, for both functionalized metal oxide, UVA-Protection Factor (UVA-PF) values were better, but UVA/UVB ratio and lambda critical value did not show a better behavior, maintaining the same values for each formulation.

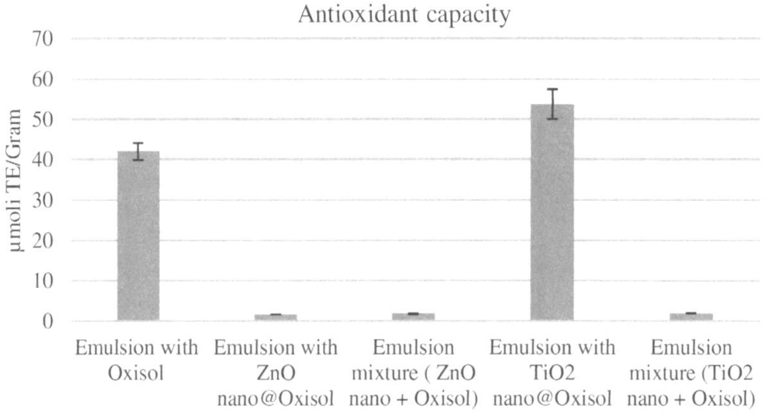

Figure 4. PCL results in formulation study for mixture and functionalized materials. Each value was obtained from three different experiments.

This boosting activity would allow half of the filter to be used to achieve the same protection, with a positive impact on the environment and in particular on marine flora and fauna. It should be remembered that there are two types of actions: the SPF (sun protection factor) value is related to the protection against UVB radiation, while UVA-PF (UVA protection factor) refers to UVA. Finally, our solution provides non-cytotoxic, non-photoreactive products that are more photoprotective than the starting materials and very potent as antioxidants, reducing sun damage on the skin.

We can therefore resume that the sustainability of a product always starts from a "a safe by design" approach. Designing according to potential toxic effects reduces the risk of generating unsuitable products, and its concept underlies the platform devised by the SbD4Nano project to share data and obtain predictions of toxicity and safety for people and the environment.
What is the challenge? Finding the best compromise, which for us is represented by the active coating with the aim of producing a finished cosmetic formulation according to the SbD rules.

6. Conclusion

Nevertheless, the growing commercial and scientific interest in the use of nanosized metal oxides in sunscreen, precise knowledge of the mechanism responsible for TiO_2 and ZnO NP toxicity is largely unknown.

With this work we demonstrated the feasibility of a SbD approach, highlighting the effectiveness and safety of NPs sunscreen formulations by the NPs coating, formulation, and the interaction of these components with UV-radiation. All information is commonly absent in the commercialized products. Furthermore, the generation of ROS could be involved even in the absence of UV radiation, so our approach permits also to avoid, anyway, the possible cyto and genotoxic occurrence.
Development of new, highly effective hybrid sunscreens (inorganic filters with polyphenols coating) is an important task to achieve both adequate photoprotection and environmental safety compromise. At the same time, it is useful to disseminate correct information and to encourage prudent use of sunscreens. It is important to continue exploring the methods that can repress the production of ROS by sunscreen NPs, rather than reducing their effects. To further minimize the risks associated with the use of NPs in sunscreens, NPs skin penetration and toxicity test guidelines should be improved and standardized. In addition, the TiO_2 and ZnO NPs physicochemical characteristics that promote skin penetration into viable skin layers must be recognized.

Acknowledgments
The authors acknowledge the European Community's Horizon 2020 Framework Program H2020 (grant number 862195 Project: SbD4Nano computing infrastructure for the definition, performance testing and implementation of safe-by-design approaches in nanotechnology supply chains; https://www.sbd4nano.eu/) to support this project.

References

[1] Watson M, Holman D M and Maguire-Eisen M 2016 Ultraviolet Radiation Exposure and Its Impact on Skin Cancer Risk *Semin. Oncol. Nurs.* 32 241–54
[2] European Commission 2009 Regulation (EC) 1223/2009 of the European Parliament and of the Council of 30 November 2009 on cosmetic products.
[3] American Academy of Dermatology Association 2018 *Memorandum: Hawaii sunscreen ban actions [press release]*
[4] Hawaii 2018 SB2571 *https//www.capitol.hawaii.gov/session2018/bills/ SB2571_.HTM.* Accessed 19 Jun 2020
[5] Miller I B, Pawlowski S, Kellermann M Y, Petersen-Thiery M, Moeller M, Nietzer S and Schupp P J 2021 Toxic effects of UV filters from sunscreens on coral reefs revisited: regulatory aspects for "reef safe" products *Environ. Sci. Eur.* 33
[6] Antoniou C, Kosmadaki M G, Stratigos A J and Katsambas A D 2008 Sunscreens - What's important to know *J. Eur. Acad. Dermatology Venereol.* 22 1110–9
[7] Schneider S L and Lim H W 2019 A review of inorganic UV filters zinc oxide and titanium dioxide *Photodermatolology, Photoimmunol. Photomed.* 35 442–6
[8] European Commission (EC) 2011 EUR-Lex - 32011H0696 - EN - EUR-Lex (Commission Recommendation of 18 October 2011 on the definition of nanomaterial) *Off. J. Eur. Union*
[9] European Commission, News articles, 10 June 2022 Directorate-General for Environment *https://environment.ec.europa.eu/news/chemicals-commission-revises-definition-nanomaterials-*

IOP Conf. Series: Materials Science and Engineering 1265 (2022) 012007 doi:10.1088/1757-899X/1265/1/012007

2022-06-10_en (accessed June 18th 2022)
[10] Koral C, Fantauzzi M, Imparato C, Papari G P, Silvestri B, Aronne A, Andreone A and Rossi A 2020 Defects in the amorphous–crystalline evolution of gel-derived TiO_2 J. Phys. Chem. C 124 23773–83
[11] Smijs T G and Pavel S 2011 Titanium dioxide and zinc oxide nanoparticles in sunscreens: Focus on their safety and effectiveness *Nanotechnol. Sci. Appl.* 4 95–112
[12] Battistin M, Dissette V, Bonetto A, Durini E, Manfredini S, Marcomini A, Casagrande E, Brunetta A, Ziosi P, Molesini S, Gavioli R, Nicoli F, Vertuani S and Baldisserotto A 2020 A New Approach to UV Protection by Direct Surface Functionalization of TiO_2 with the Antioxidant Polyphenol Dihydroxyphenyl Benzimidazole Carboxylic Acid *Nanomaterials* 10, 231
[13] Battistin M, Durini E, Dissette V, Bonetto A, Marcomini A, Casagrande E, Brunetta A, Ziosi P, Molesini S, Gavioli R, Nicoli F, Manfredini S, Vertuani S and Baldisserotto A 2019 Synthesis and characterization of new multifunctional self-boosted filters for UV protection: ZnO complex with dihydroxyphenyl benzimidazole carboxylic acid *Molecules* 24, 4546
[14] van Driel B A, Kooyman P J, van den Berg K J, Schmidt-Ott A and Dik J 2016 A quick assessment of the photocatalytic activity of TiO_2 pigments - From lab to conservation studio! *Microchem. J.* 126 162–71
[15] Dimitrovska Cvetkovska A, Manfredini S, Ziosi P, Molesini S, Dissette V, Magri I, Scapoli C, Carrieri A, Durini E and Vertuani S 2017 Factors affecting SPF in vitro measurement and correlation with in vivo results *Int. J. Cosmet. Sci.* 39 310–9
[16] Popov I and Lewin G 1999 Antioxidative homeostasis: Characterization by means of chemiluminescent technique *Methods Enzymol.* 300 437–56
[17] Lewin G and Popov I 1994 Photocheluminescent detection of antiradical activity; III: a simple assay of ascorbate in blood plasma *J. Biochem. Biophys. Methods* 28 277–82

IOP Conf. Series: Materials Science and Engineering 1265 (2022) 012008 doi:10.1088/1757-899X/1265/1/012008

Characterization of Disposable Facemasks for COVID-19 Through Colorimetric Analysis

E. Fraschetti[1], D. Puglisi[2], G. Domènech-Gil[2], A. Buzzin[1], A. Mastrandrea[1], I. Mazzetta[1], G. de Cesare[1], S. Casalinuovo[1], S. Quaranta[3], and D. Caputo[1]

[1]Department of Information Engineering, Electronics and Telecommunications, Sapienza University of Rome, Italy
[2]Department of Physics, Chemistry and Biology, Sensor and Actuator System Division, Linköping University, Sweden
[3]Consiglio Nazionale delle Ricerche, Institute for the Study of Nanostructured Materials (CNR ISMN), Rome, Italy

domenico.caputo@uniroma1.it

Abstract. Many aspects of the world population's daily life have been recently changed by the events following the SARS-COV-2 pandemic outbreak. Among all the consequences, wearing face masks has become a common routine to protect from virus transmission risks. This work presents a simple colorimetric system able to detect the carbon dioxide (CO_2) saturation inside a disposable face mask, which is useful to determine the level of wear and degradation and to visually provide indications on its disposal time. The experiments were carried out by wearing a FFP2 face mask externally treated with a phenolphthalein solution and including in its breathing zone a CO_2 sensor. Changes in face mask color were recorded by a camera and analyzed with ImageJ. A strong correspondence was found between the high values of CO_2 detected by the sensor and the analyzed data. The results are promising and suggest further efforts in developing easy-to-use colorimetric methods as a visual indicator of the life cycle of a disposable face mask.

1. Introduction

In March 2020, the rapid spreading of the SARS-COV-2 virus caused the declaration of the pandemic state by the World Health Organization (WHO) [1]. This outbreak forced the general population to change their daily habits, adopting strict emergency measures such as social distancing, remote working, and use of face masks as protection from virus transmission risks [2, 3]. Being considered as a new accepted routine [4], the practice of wearing a face mask has been instrumental in hindering the spread of the virus [5] but, at the same time, introduced potential health risks when prolonged usage without a correct replacement is practiced on regular basis [6, 7, 8].

In this framework, we present a simple colorimetric system to predict the level of deterioration of disposable face masks based on a phenolphthalein acid-base indicator, and suitable for quantifying carbon dioxide (CO_2) concentration in its interior. A study was performed by coloring the masks with a phenolphthalein solution and observing color changes during usage in breathing experiments. At the same time, concentration of CO_2 inside the masks was monitored by means of a miniaturized sensor. The relation between the color change and CO_2 concentration has been studied through an RGB analysis, experimentally highlighting the progressive decrease of red components in the mask's color as its internal environment maintains its exposition to high levels of CO_2 content for prolonged time.

2. Experimental setup

A SCD30 sensor (from Sensirion AG, Switzerland) was used to monitor CO_2 in real time, and was secured inside a FFP2 face mask [9]. Based on the NDIR technology, the CO_2 sensor has a declared accuracy of ± (30 ppm + 3%) over a measurement range of 400 – 10.000 ppm and a repeatability of ± 10 ppm. The sensor was connected to a Raspberry Pi 4 Model B to process the acquired data. The Raspberry processor was connected to a Raspberry Pi Camera Module to take pictures during the experiment. Figure 1 depicts the experimental setup.

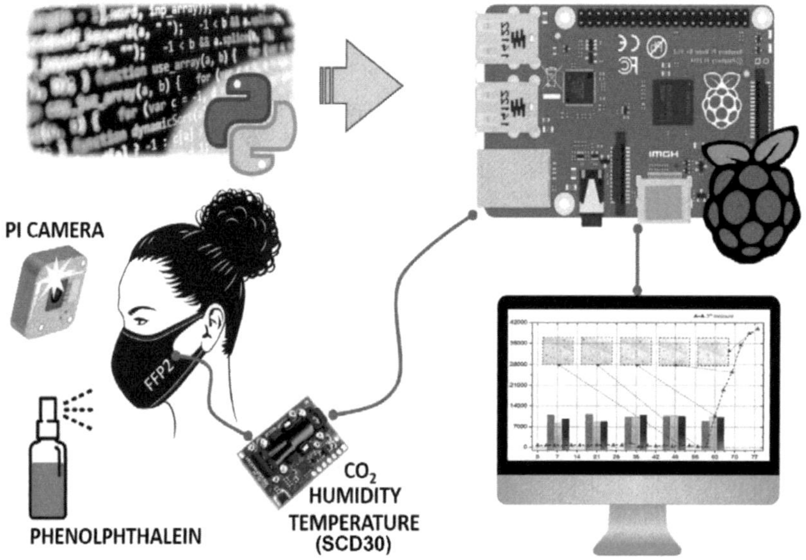

Figure 1. Simultaneous CO_2 monitoring inside a face mask via colorimetric and standard sensing methods: Experimental setup.

A phenolphthalein solution was prepared to perform the colorimetric analysis. Phenolphthalein ($C_{20}H_{14}O_4$) is a common pH indicator used in acid-base titrations. Phenolphthalein is an odourless white crystalline powder at room temperature, sparingly soluble in water but soluble in ethanol. It is not particularly irritating to eyes and skin, and it is not known to have sensitising effects. It is toxic if ingested, although it was often used in pharmaceutical products for its laxative properties in the past. From a chemical point of view, phenolphthalein is a very weak acid and an accurate pH indicator. It is colourless at pH values below 8, while at pH values above 9 the hydroxyls lose their protons, and the molecule imparts an intense purple colour to the solution [10].

The colorimetric solution was prepared using 100 mL of distilled water and 0.1 g of phenolphthalein, with 7 mL of Triethanolamine (TEA) added as a pH balancer. With this ratio, the solution appeared purple in colour, as visible in Figure 2.

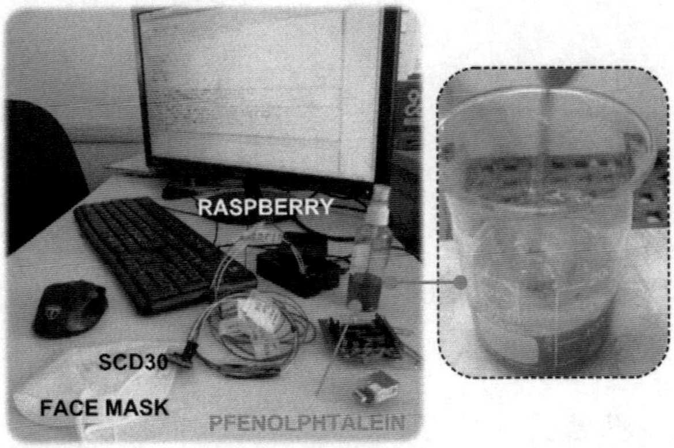

Figure 2. Picture of the experimental setup, with a phenolphthalein solution in a spray container on the right.

The tests were carried out as follows:
1. spraying the phenolphthalein solution on the FFP2 face mask with the SCD30 inside the breathing environment.
2. wearing the mask and performing 300 s of normal and relaxed breathing.

The amount of CO_2 was monitored by the SCD30 and by a visual colorimetric analysis, simultaneously. The results were then compared to address the congruency of the colorimetric method for this kind of applications.

3. Results

The SCD30 data highlighted, as a first result, a rapid increase of the CO_2 presence inside the mask environment from the beginning of the test. Secondly, the detected CO_2 oscillates inside a range from 10000 ppm to 30000 ppm, behaving in accordance with the wearer's breath (Figure 3a).

To confirm the high amounts of CO_2 inside the breathing environment, five pictures of the same facemask's spot were taken, the first one at t = 0 and four more every 75 seconds. The colour variations of the collected images were analysed in detail by RGB (Red, Green and Blue) analysis, carried out using ImageJ (developed by the US National Institutes of Health) [11]. The same region of interest was investigated in every image, and the RGB data were extracted from the region studied. The mean values in the five images were evaluated for every colour component to suppress any noise due to involuntary movements and variations in the environment's brightness during the experiment. Subsequently, the red component (R) was isolated and normalized to 100.

Figure 3b shows the time evolution of the R component, with the corresponding pictures over the plot: the color change from bright pink to white, and the subsequent decrease of red content in the pictures, indicates that the sprayed phenolphthalein solution was continuously exposed to high levels of CO_2, confirming the sensor data (Figure 3a).

Comparison between figure 3a and 3b demonstrates that continuously breathing in a FFP2 facemask quickly brings to breathing in a CO2-saturated environment; this slowly modifies the color of the mask, to provide the wearer with a visual indication of how long he is breathing inside a CO_2-saturated environment.

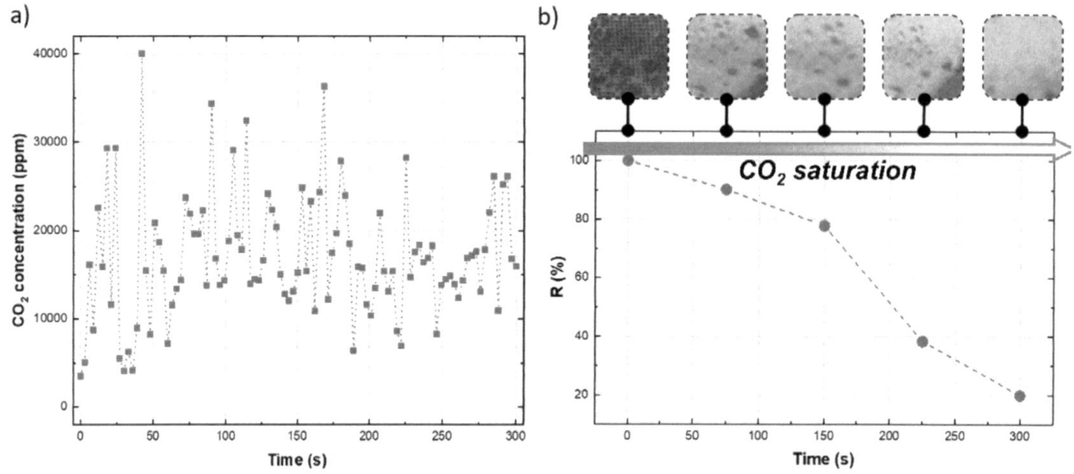

Figure 3. a) CO_2 concentration as monitored by the SCD30 sensor during the experiment. b) Red component of the 5 pictures taken by the camera and elaborated by ImageJ.

4. Conclusions

This work presents a simple colorimetric system able to detect the CO_2 saturation inside a disposable face mask. This aim has been achieved by simultaneously monitoring the CO_2 concentration by a commercial sensor and the color change of the mask externally sprayed with a phenolphthalein solution. Time evolution of mask images through RGB analysis highlights a progressive decrease of the red component as the internal mask environment maintains its exposition to high CO_2 concentration.

The reported experiments demonstrate for the first time the relationship between the exposition to high CO_2 concentration for prolonged time and the face mask visual degradation through color change. Moreover, such analysis may be exploited as a simple method, coupled to macro/micro inspection of the mask, to relate CO_2 concentration to the level of deterioration (dirtying and pore clogging) of the mask levels and its possible side effects (e.g., headaches, fatigue, loss of concentration) related to prolonged use of masks [12]. The presented results are promising and suggest further efforts in developing easy-to-use colorimetric methods as a visual indicator of the life cycle of a disposable face mask.

References

[1] Ghebreyesus, T. A. 2020 "World Health Organization. WHO Director-General's opening remarks at the media briefing on COVID-19" 25 May 2020.

[2] Tso, R. V., and Cowling, B. J. 2020 "Importance of face masks for COVID-19: A call for effective public education." *Clinical Infectious Diseases* **71** pp 2095-2198.

[3] Wong, S. H., Teoh, J. Y., Leung, C. H., Wu, W. K., Yip, B. H., Wong, M. C., and Hui, D. S. 2020 "COVID-19 and public interest in face mask use." *American Journal of Respiratory and Critical Care Medicine* **202** pp 453-455.

[4] Rab, S., Javaid, M., Haleem, A. and Vaishya, R. 2020 "Face masks are new normal after COVID-19 pandemic." *Diabetes and Metabolic Syndrome: Clinical Research and Reviews* **14** pp 1617–1619.

[5] Leung, N. H., Chu, D. K., Shiu, E. Y., Chan, K. H., McDevitt, J. J., Hau, B. J., and Cowling, B. J. 2020 "Respiratory virus shedding in exhaled breath and efficacy of face masks." *Nature medicine* **26** pp 676-680.

[6] Lee, L. Y. K., Lam, E. P. W., Chan, C. K., Chan, S. Y., Chiu, M. K., Chong, W. H., and Wu, C. W. 2020 "Practice and technique of using face mask amongst adults in the community: a cross-sectional descriptive study." *BMC Public Health* **20** pp 1-11.

[7] Li, Y., Tokura, H., Guo, Y. P., Wong, A. S. W., Wong, T., Chung, J., and Newton, E. 2005 "Effects of wearing N95 and surgical facemasks on heart rate, thermal stress and subjective sensations." *International archives of occupational and environmental health* **78** pp 501-509.

[8] Yang, Q., Li, H., Shen, S., Zhang, G., Huang, R., Feng, Y., and Ma, S. 2018 "Study of the micro-climate and bacterial distribution in the deadspace of N95 filtering face respirators." *Scientific reports* **8**, pp 1-13.

[9] Rengasamy, S., Shaffer, R., Williams, B. and Smit, S. A. 2017 "Comparison of facemask and respirator filtration test methods." *J. Occup. Environ. Hyg.* **14** pp 92–103.

[10] Liu, Z., Luo, F., and Chen, T. 2004 "Phenolphthalein immobilized membrane for an optical pH sensor." *Analytica chimica acta* **510** pp 189-194.

[11] R Abràmoff, M. D., Magalhães, P. J., and Ram, S. J. 2004 "Image processing with ImageJ." *Biophotonics international* **11** pp 36-42.

[12] Geiss, O. 2021 "Effect of wearing face masks on the carbon dioxide concentration in the breathing zone." *Aerosol and Air Quality Research* **21** p 200403.

IOP Conf. Series: Materials Science and Engineering 1265 (2022) 012009 doi:10.1088/1757-899X/1265/1/012009

Waste valorization by nanotechnology approaches for sustainable crop protection: a mini review

D Schiavi[1, *], **V Di Lorenzo**[1], **S Francesconi**[1], **S Giovagnoli**[2], **E Camaioni**[2] **and G M Balestra**[1]

[1] Department of Agriculture and Forest Sciences (DAFNE), University of Tuscia, Via S. Camillo de Lellis snc, 01100, Viterbo, Italy
[2] Department of Pharmaceutical Sciences (DSF), University of Perugia, Via del Liceo 1, 06123, Perugia, Italy

[*] schiavi@unitus.it

Abstract. Agri-food waste management represent for several supply-chains a pending issue, since most of the lignocellulosic biomasses produced by the agricultural practices are often intended for energy production, due to their high disposal costs. Nanotechnology could represent a sustainable way to valorise agricultural waste by extracting highly performing nanomaterials and reusing them as nanocarriers or nanopesticides to protect crops. Cellulose and lignin, which are the main component of plant residues display antimicrobial properties when reduced to their nanosized forms, being able to act as innovative tools to counteract plant pathogens, both bacterial and fungal ones. In particularly cellulose can be extracted from different matrices to obtain cellulose nanocrystals (CNC), while lignin is often reduced in nanoparticles (LNP). In this mini review the current and most recent results in terms of antimicrobial activities against plant pathogens by waste-based lignocellulosic nanomaterial are showed.

1. Introduction

Nanotechnology can open new fascinating scenarios especially in those disciplines where finding sustainable alternative in industrial processes is hugely required nowadays. In the last years the application of nanotechnology in agriculture has showed to be a very promising tool for making crops management keener and greener, from plant nurseries to post-harvest processes. Nanotechnology can be successfully applied in crop nutrition, plant protection, seed priming, biosensoring, soil and water remediation and smart packaging [1]. Several works and researches already showed the potential of nanotechnology and nanomaterials for plant management [2,3]. Nanomaterials can be obtained from two different main approaches: the *top-down* and the *bottom-up* one. The first approach defines nanomaterials' synthesis as a reduction of a bulk material to its nanosized form, while the latter considers the synthesis of a nanomaterial as an assembling of smaller molecules up to the nanosized final dimension. Both approaches can be successfully used to obtain nanomaterials showing new biological properties in comparison with their primary counterparts [4]. Through several methods it is possible to synthetize nanomaterials even from agricultural waste biomasses.

Indeed lignocellulosic waste disposal plays an important role in the sustainability of the whole agri-food sector, since a global production of more than 24 million tons is estimated every year [5]. Lignocellulosic biomasses are produced in many steps of the crop management, such as pruning, shelling, harvesting and transplanting, and in most of these cases they are intended for energy production. While many researches have already pointed out the possibility of using agricultural waste as a source for extracting active ingredients, such as phenols and essentials oils, in comparison there are just few works that highlight the use of lignocellulosic biomasses as a bulk for the synthesis of organic nanomaterials [6]. Among them, the ones derived from cellulose and lignin, are grabbing the attention of the scientific community in the last few years, since their use can be extended to different application fields. Lignocellulosic nanomaterials can be exploited as nanocarriers by encapsulating active principles or linking them through surface modifications; as reinforcement phase in packaging materials, or as highly performing additives in industrial processes [7,8]. Even biomedical applications are being considered since the lack of cytotoxicity and the scientific evidences about their biocompatibility [9–11]. These nanomaterials find a potential use in agriculture too, by improving plant nutrition and protection. In this scenario it is fundamental to remember that plant protection strategies nowadays still rely on the use of copper-based compounds and synthetic agrochemicals in most cases [12]. Finding alternative pesticides is mandatory in the complex situation of the lack of resources, climate changes, pesticide resistance, the growing of organic agriculture and the consumers' awareness awakening on the need of more sustainable supply-chains.

2. Lignocellulosic-based nanomaterials for plant protection

The recent advances in biopolymers science have led to the possibility of developing more functional nanomaterials based on lignocellulosic compounds, exploiting the huge amount of agri-food waste produced by the industry. Lignocellulosic nanomaterials are currently being studied for several applications in agriculture since their low cost, abundance, and biocompatibility. These unique features can be paired with the advantages provided by nanotechnology, which grant interesting biochemical properties to cellulose and lignin when reduced to nanosized forms. Thus lignocellulosic nanomaterials were investigated for the capability of acting as nanocarriers for other molecules, such as fertilizers or pesticides, additives in smart packaging, soil and water remediation tools [13–15]. Notably lignocellulosic nanomaterials can be exploited also as antimicrobial compounds themselves, which open wide scenarios in defining sustainable crop protection strategies [16]. In the last years huge improvements were performed in studying the antimicrobial mechanisms of lignocellulosic polymers respect to bacteria and fungi, even if most of these researches were focused on human-related pathogens [17]. Indeed, just few works reported lignin and cellulose antimicrobial activities on plant pathogens, especially in their nanoform (Table 1). However, these results cannot be ruled out, since their promising potential for crop protection [18].

Table 1. Antimicrobial properties of lignocellulosic nanomaterials.

Nanomaterial	Source	Synthesis method	Application	Mode of action	Reference
CNC	Cotton pulp	Acid hydrolysis	*P. aeruginosa, P. fluorescens*	Flocculation, depletion	Sun *et al.* (2012); Sun *et al.* (2014)
	Elephant grass	TEMPO-oxidation	*E. coli*	Membrane integrity's loss	Noronha *et al.* (2021)
	Microcrystalline cellulose	Acid hydrolysis	*P. syringae pv. tomato*	Growth inhibition	Yang *et al.* (2016)
	Hazelnut shells and prunings	Acid hydrolysis	*X. arboricola pv. corylina*	Growth inhibition	Schiavi *et al.* (2022)
	Olive tree prunings	Acid hydrolysis	*P. savastanoi pv. savastanoi*	Growth and biofilm inhibition	Schiavi *et al.* (2022)

Nanomaterial	Source	Synthesis method	Application	Mode of action	Reference
LNP	Bamboo chips	Organosolv fractionaction	*E. coli, S. aureus*	Growth inhibition	Yun *et al.* (2021)
	Giant reed	Hydrochloric acidolysis	*P. syringae pv. tomato*	Growth inhibition	Yang *et al.* (2016)
	Giant reed	Hydrochloric acidolysis	*X. arboricola pv. pruni, P. carotovorum subsp. odoriferum*	Growth inhibition	Yang *et al.* (2016)
	Hazelnut shells and prunings	Solvent-antisolvent method	*X. arboricola pv. corylina*	Growth inhibition	Schiavi *et al.* (2022)
	Beech sawdust	Antisolvent method	*P. cactorum*	Growth inhibition	Vettraino *et al.* (2022)

2.1. Cellulose nanocrystals
Cellulose is the most abundant organic polymer on Earth, since its biosynthesis can be performed by plants, bacteria, and other microorganisms. Cellulose' structure consists of repeated units of cellobiose, a dimer of β1-4 linked glucose. These molecules rearrange themselves as long fibrils granting elasticity and mechanic resistance to plant cell walls. Cellulose nanocrystals (CNC), which represent the most investigated nanosized form of cellulose, are obtained through several protocols, such as acid hydrolysis or TEMPO-mediated oxidation followed by defibrillation; several alternative methods based on mechanical treatments were proposed too. CNC are characterized by an acicular shape, with the main dimension of length of 100-500 nm, while width ranges often between 10 and 30 nm. CNC have showed some peculiar mechanical properties like a high elastic modulus, a low thermal expansion coefficient and a low density (Figure 1) [19–21]. Moreover CNC have been studied for the possibility of exploiting their abundant hydroxyl groups as sites for surface modifications [22]. Chemical modifications have led to the use of CNC as carrier for drugs and active molecules in biomedical, pharmaceutical, and agricultural fields. Since the huge interest regarding the application of nanocrystalline cellulose, several biomasses were investigated as alternative and low-cost source of cellulose. CNC were successfully obtained from pruning residues of kiwifruit plants, sunflowers stalks, hemp fibers, walnut shells. Their major application was proposed as reinforcement phase in packaging materials or in innovative biopolymers [23–26]. Although the many researches in this sector, only few works have described the potential antimicrobial properties of nanocellulose. Among the antibacterial mechanisms of CNC, a depletion effect were recorded on *P. aeruginosa* and *P. fluorescens*, that could have led, as authors suggested, to the flocculation of cells inhibiting their adhesion capacity [27,28]. Another CNC modes of action against bacteria were proposed by Noronha *et al.* (2021), which have observed a loss in the membrane's integrity in *E. coli* cells when exposed to CNC [29]. In crop protection some interesting results were achieved too. CNC obtained from microcrystalline cellulose showed a promising inhibitory effect on *Pseudomonas syringae* pv. *tomato* (Pst), the causal agent of tomato bacterial speck, when combined with lignin and poly (lactic acid) in ternary films [30]. CNC-based microparticles of chitosan revealed the capability of inhibiting the *in vitro* growth of Pst as well as the reduction of disease incidence and severity when applied on tomato seedlings, in a comparable way to copper [31]. Currently, only few works have reported antimicrobial activity of waste-based CNC against phytopathogenic bacteria. In the first one, the antibacterial activity of CNC extracted by hazelnut shells and pruning residues were observed by a microdilution assay against *Xanthomonas arboricola* pv. *corylina* (Xac), the causal agent of hazelnut bacterial

blight. At the same time hazelnut seedlings treated with 1% w/v CNC suspension were reported to develop a less heavy symptomatic display after being artificially inoculated with Xac [32]. Another recent research, proposed by Schiavi *et al.* (2022), has highlighted the antimicrobial activity of CNC obtained from olive tree pruning residues through a chemical bleaching followed by an acid hydrolysis, against *Pseudomonas savastanoi* pv. *savastanoi* (Psav), the olive knot disease's causal agent. Reported results showed the ability of 1% w/v CNC suspension to *in vitro* inhibit the growth of Psav as well as the biofilm production. This properties were confirmed also on artificially-inoculated olive tree seedlings, where a reduction of leaf epiphytic survival of the bacteria was observed, in a comparable way to the one showed by copper sulphate at the field dose [33]. Although the promising results, many more studies are required to completely understand the role of nanocrystalline cellulose in interactions with plant pathogenic microorganisms.

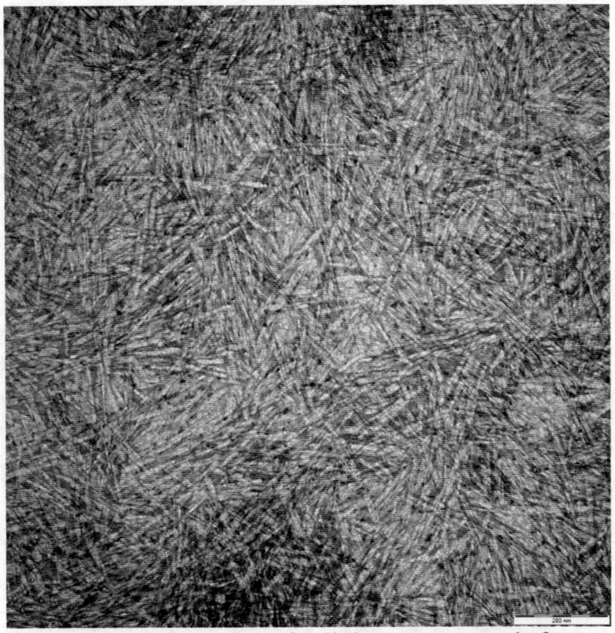

Figure 1. TEM image of cellulose nanocrystals from hazelnut pruning residues (Authors' unpublished data).

2.2. Lignin nanoparticles

Lignin is the second most abundant natural polymer on Earth since it can be found in the cell walls of higher plants. Its presence in tissues grants stability and resistance providing at the same time to a natural barrier against pathogens. Antimicrobial and antioxidant properties of lignin are due to its phenolic-based composition: lignin is in fact a very complex natural polymer, whose structure it has not been fully identified yet. It is mainly composed of p-coumaryl alcohol, sinapyl alcohol, coniferyl alcohol, and several aromatic subunits [34]. Their ratio and the presence of other phenolic functional groups together with the molecular weight, differ from species to species and among the same plants also considering the phenological phase and the phytosanitary status. These differences could affect the final size, the molecular weight and the thermal stability of lignin nanoproducts [35]. For many time lignin has represented a subproduct of the wood pulp krafting industrial process, but it can be naturally found in order of 15-45% in all the agricultural biomasses. Extraction and isolation of lignin from agricultural wastes, as well as synthesis of lignin nanoparticles (LNP) or nanospheres, can be achieved through different methods (Figure 2). The most diffuses ones exploit solvent exchange protocols, precipitation, ultrasonication or even the self-assembling properties of lignin [36]. Many researches have reported LNP coming from agrowastes, such as chestnut, acacia or bamboo, to possess antimicrobial properties, especially against bacteria [37–39]. These mechanisms seem to be mostly related to LNP antioxidant properties, even if several modes of action were supposed. Among those there is the possibility that LNP can induce oxidative stress in cells by accumulating reactive oxygen species (ROS) on their surface and release them when in contact with biological membranes. Moreover, due to their small size LNP can penetrate in the living cells causing ATP depletion or changes in pH inner values, leading cells to death as a results [40]. Lignin was successfully obtained in its nanosized forms from the residues of *Arundo donax* used in bioethanol production. These nanoparticles showed interesting antibacterial properties against Pst, when combined in a ternary polymeric films with poly (lactic acid) and CNC [30]. The same LNP were used together with chitosan and polyvinyl alcohol to realize an effective film produced through the solvent casting method to counteract *Xanthomonas arboricola* pv. *pruni* (Xap) and *Pectobacterium carotovorum* subsp. *odoriferum* (Pco), the causal agents of stone fruits canker and vegetables soft rot respectively. The experiments revealed the capability to reduce bacterial cells multiplication through time and its utilization was suggested to prevent infections in post-harvest products [41]. LNP were successfully obtained through a solvent-antisolvent method from hazelnut shelling and pruning residues and successfully employed against Xac, showing inhibition properties on the *in vitro* bacterial growth as well as the capability of reducing the disease incidence on artificially inoculated hazelnut plants, in a comparable way to copper oxychloride [32]. LNP can be successfully used as nanocarriers for active molecules too. Lignin capsules can be obtained by different methods as reported in literature: entrapment, encapsulation or adsorption. Moreover the process of encapsulation grants additional thermal stability, UV-protection and controlled release behavior to the active molecules [42]. This possibility was widely investigated by using kraft lignin, but now several researches have highlighted the effective use of pristine lignin too. Zikeli *et al.* (2020) were able to use beech sawdust as a source of LNP for entrapping essential oils derived from cinnamon and thyme, study their stability and release, suggesting their potential use as biocide nanocarriers [43]. This intuition was actually verified in the work proposed by Vettraino *et al.* (2022), in which the same LNP loaded with thyme essential oil had demonstrated the capability of reducing the mycelium growth of *Phytophthora cactorum*, as well as its mortality on black pine seedlings [44].

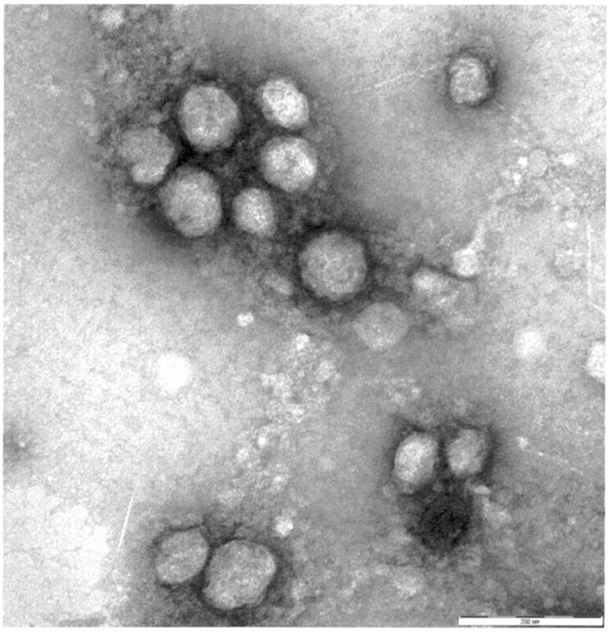

Figure 2. TEM image of lignin nanoparticles from hazelnut pruning residues (Authors' unpublished data).

3. Conclusion

Since the urge of a greener agriculture is everyday more pressing, plant protection strategies need innovative tools to counteract the impact of climate changes and at the same time the diminishing of traditional active principles efficacy. One of the best alternatives is nowadays represented by nanotechnology. Nanomaterials show different antimicrobial mechanisms, acting directly on the pathogens or as plant resistance elicitors, or even as nanocarriers for active molecules. Nanotechnology also represents a valid alternative for agri-food waste valorization, since lignocellulosic biomasses can be used as source for cellulose or lignin-based nanomaterials. Cellulose and lignin in their several nanometric forms (crystals and fibrils and particles) are currently under the scientific community's lens, due to their unique interesting microbiological properties together with their biocompatibility, although many researches have to be developed to better understand how transfer these preliminary data from lab to field, especially referring to the distribution of these compounds. In this sense there is still a huge lack in understanding the chemical stability of the nanomaterials, their fate in soil, as well as their solubility, the proper concentration and distribution mode to maintain the antimicrobial properties. Nevertheless, the use of lignocellulosic-based nanomaterials could provide an effective solution to two main challenges of modern agriculture: a new way to boost agricultural waste management while at the same time providing sustainable tools to control plant pathogens, by a circular economy approach.

Acknowledgments

This works was funded by MUR (Italian Ministry for University and Research) (Law 232/2016, Departments of Excellence) and PON Ricerca e Innovazione 2014–2020: Nanotecnologie chimiche green per la protezione sostenibile delle piante (NEMESI) (ARS01_01002 CUP: F36C1800018000X).

References

[1] Prasad R, Bhattacharyya A and Nguyen Q D 2017 Nanotechnology in sustainable agriculture: Recent developments, challenges, and perspectives *Front. Microbiol.* **8** 1014

[2] Ojha S, Singh D, Sett A, Chetia H, Kabiraj D and Bora U 2018 Nanotechnology in Crop Protection *Nanomaterials in Plants, Algae, and Microorganisms* vol 1 pp 345–91

[3] Balestra G M and Fortunati E 2022 *Nanotechnology-based sustainable alternatives for the managements of plant diseases* ed G M Balestra and E Fortunati (Amsterdam, The Netherlands: Elsevier)

[4] Kumar R, Dalvi S V. and Siril P F 2020 Nanoparticle-Based Drugs and Formulations: Current Status and Emerging Applications *ACS Appl. Nano Mater.* **3** 4944–61

[5] Fortunati E, Mazzaglia A and Balestra G M 2019 Sustainable control strategies for plant protection and food packaging sectors by natural substances and novel nanotechnological approaches *J. Sci. Food Agric.* **99** 986–1000

[6] García A, Gandini A, Labidi J, Belgacem N and Bras J 2016 Industrial and crop wastes: A new source for nanocellulose biorefinery *Ind. Crops Prod.* **93** 26–38

[7] Tribot A, Amer G, Abdou Alio M, de Baynast H, Delattre C, Pons A, Mathias J D, Callois J M, Vial C, Michaud P and Dussap C G 2019 Wood-lignin: Supply, extraction processes and use as bio-based material *Eur. Polym. J.* **112** 228–40

[8] Puglia D, Pezzolla D, Gigliotti G, Torre L, Bartucca M L and Del Buono D 2021 The opportunity of valorizing agricultural waste, through its conversion into biostimulants, biofertilizers, and biopolymers *Sustain.* **13** 2710

[9] Roman M 2015 Toxicity of cellulose nanocrystals: A review *Ind. Biotechnol.* **11** 25–33

[10] Hassan G, Forsman N, Wan X, Keurulainen L, Bimbo L M, Stehl S, Van Charante F, Chrubasik M, Prakash A S, Johansson L S, Mullen D C, Johnston B F, Zimmermann R, Werner C, Yli-Kauhaluoma J, Coenye T, Saris P E J, Österberg M and Moreira V M 2020 Non-leaching, Highly Biocompatible Nanocellulose Surfaces That Efficiently Resist Fouling by Bacteria in an Artificial Dermis Model *ACS Appl. Bio Mater.* **3** 4095–108

[11] Lin N and Dufresne A 2014 Nanocellulose in biomedicine: Current status and future prospect *Eur. Polym. J.* **59** 302–25
[12] La Torre A, Iovino V and Caradonia F 2018 Copper in plant protection : current situation and prospects *Phytopathol. Mediterr.* **57** 201–36
[13] Maćczak P, Kaczmarek H and Ziegler-Borowska M 2020 Recent achievements in polymer bio-based flocculants for water treatment *Materials (Basel).* **13** 3951
[14] Ten E and Vermerris W 2015 Recent developments in polymers derived from industrial lignin *J. Appl. Polym. Sci.* **132** 42069
[15] Lima P H C de, Antunes D R, Forini M M de L, Pontes M da S, Mattos B D and Grillo R 2021 Recent Advances on Lignocellulosic-Based Nanopesticides for Agricultural Applications *Front. Nanotechnol.* **3** 809329
[16] Fortunati E and Balestra G M 2019 Lignocellulosic materials as novel carriers, also at nanoscale, of organic active principles for agri-food applications *Biomass, Biopolymer-Based Materials, and Bioenergy: Construction, Biomedical, and other Industrial Applications* ed V Deepak, E Fortunati, S Jain and X Zhang (Amsterdam, The Netherlands: Elsevier Ltd) pp 161–78
[17] Lobo F C M, Franco A R, Fernandes E M and Reis R L 2021 An overview of the antimicrobial properties of lignocellulosic materials *Molecules* **26** 1749
[18] Di Lorenzo V, Schiavi D, Petrucci C and Balestra G M 2021 Organic Nanoparticles from Lignocellulosic Agricultural Residues for Plant Protection Applications *AIP Conf. Proc.* **2416** 020013
[19] Kamel R, El-Wakil N A, Dufresne A and Elkasabgy N A 2020 Nanocellulose: From an agricultural waste to a valuable pharmaceutical ingredient *Int. J. Biol. Macromol.* **163** 1579–90
[20] Vanderfleet O M, Reid M S, Bras J, Heux L, Godoy-Vargas J, Panga M K R and Cranston E D 2019 Insight into thermal stability of cellulose nanocrystals from new hydrolysis methods with acid blends *Cellulose* **26** 507–28
[21] Brinchi L, Cotana F, Fortunati E and Kenny J M 2013 Production of nanocrystalline cellulose from lignocellulosic biomass : Technology and applications *Carbohydr. Polym.* **94** 154–69
[22] Habibi Y 2014 Key advances in the chemical modification of nanocelluloses *Chem. Soc. Rev.*
[23] Luzi F, Fortunati E, Giovanale G, Mazzaglia A, Torre L and Balestra G M 2017 Cellulose nanocrystals from Actinidia deliciosa pruning residues combined with carvacrol in PVA_CH films with antioxidant/antimicrobial properties for packaging applications *Int. J. Biol. Macromol.* **104** 43–55
[24] Fortunati E, Luzi F, Jiménez A, Gopakumar D A, Puglia D, Thomas S, Kenny J M, Chiralt A and Torre L 2016 Revalorization of sunflower stalks as novel sources of cellulose nanofibrils and nanocrystals and their effect on wheat gluten bionanocomposite properties *Carbohydr. Polym.* **149** 357–68
[25] Hemmati F, Jafari S M, Kashaninejad M and Barani Motlagh M 2018 Synthesis and characterization of cellulose nanocrystals derived from walnut shell agricultural residues *Int. J. Biol. Macromol.* **120** 1216–24
[26] Luzi F, Fortunati E, Jiménez A, Puglia D, Pezzolla D, Gigliotti G, Kenny J M, Chiralt A and Torre L 2016 Production and characterization of PLA_PBS biodegradable blends reinforced with cellulose nanocrystals extracted from hemp fibres *Ind. Crops Prod.* **93** 276–89
[27] Sun X, Danumah C, Liu Y and Boluk Y 2012 Flocculation of bacteria by depletion interactions due to rod-shaped cellulose nanocrystals *Chem. Eng. J.* **198–199** 476–81
[28] Sun X, Lu Q, Boluk Y and Liu Y 2014 The impact of cellulose nanocrystals on the aggregation and initial adhesion of Pseudomonas fluorescens bacteria *Soft Matter* **10** 8923–31
[29] Noronha V T, Camargos C H M, Jackson J C, Souza Filho A G, Paula A J, Rezende C A and Faria A F 2021 Physical membrane-stress-mediated antimicrobial properties of cellulose nanocrystals *ACS Sustain. Chem. Eng.* **9** 3203–12
[30] Yang W, Fortunati E, Dominici F, Giovanale G, Mazzaglia A, Balestra G M, Kenny J M and

Puglia D 2016 Synergic effect of cellulose and lignin nanostructures in PLA based systems for food antibacterial packaging *Eur. Polym. J.* **79** 1–12
[31] Schiavi D, Balbi R, Giovagnoli S, Camaioni E, Botticella E, Sestili F and Balestra G M 2021 A green nanostructured pesticide to control tomato bacterial speck disease *Nanomaterials* **11** 1852
[32] Schiavi D, Ronchetti R, Lorenzo V Di, Salustri M, Petrucci C, Vivani R, Giovagnoli S, Camaioni E and Balestra G M 2022 Circular Hazelnut Protection by Lignocellulosic Waste Valorization for Nanopesticides Development *Appl. Sci.* **12** 2604
[33] Schiavi D, Francesconi S, Taddei A R, Fortunati E and Balestra G M 2022 Exploring cellulose nanocrystals obtained from olive tree wastes as sustainable crop protection tool against bacterial diseases *Sci. Rep.* **12** 6149
[34] Liao J J, Latif N H A, Trache D, Brosse N and Hussin M H 2020 Current advancement on the isolation, characterization and application of lignin *Int. J. Biol. Macromol.* **162** 985–1024
[35] Jiang Z, Ma Y, Guo X, Remón J, Tsang D C W, Hu C and Shi B 2021 Sustainable production of lignin micro-/nano-particles (LMNPs) from biomass: Influence of the type of biomass on their self-assembly capability and physicochemical properties *J. Hazard. Mater.* **403** 123701
[36] Duarah P, Haldar D and Purkait M K 2020 Technological advancement in the synthesis and applications of lignin-based nanoparticles derived from agro-industrial waste residues: A review *Int. J. Biol. Macromol.* **163** 1828–43
[37] Aadil K R, Pandey N, Mussatto S I and Jha H 2019 Green synthesis of silver nanoparticles using acacia lignin , their cytotoxicity , catalytic , metal ion sensing capability and antibacterial activity *J. Environ. Chem. Eng.* **7** 103296
[38] Zikeli F, Vinciguerra V, Taddei A R, D'Annibale A, Romagnoli M and Mugnozza G S 2018 Isolation and characterization of lignin from beech wood and chestnut sawdust for the preparation of lignin nanoparticles (LNPs) from wood industry side-streams *Holzforschung* **72** 961–72
[39] Yun J, Wei L, Li W, Gong D, Qin H, Feng X, Li G, Ling Z, Wang P and Yin B 2021 Isolating High Antimicrobial Ability Lignin From Bamboo Kraft Lignin by Organosolv Fractionation *Front. Bioeng. Biotechnol.* **9** 683796
[40] Yang W, Fortunati E, Gao D, Balestra G M, Giovanale G, He X, Torre L, Kenny J M and Puglia D 2018 Valorization of Acid Isolated High Yield Lignin Nanoparticles as Innovative Antioxidant/Antimicrobial Organic Materials *ACS Sustain. Chem. Eng.* **6** 3502–14
[41] Yang W, Owczarek J S, Fortunati E, Kozanecki M, Mazzaglia A, Balestra G M, Kenny J M, Torre L and Puglia D 2016 Antioxidant and antibacterial lignin nanoparticles in polyvinyl alcohol/chitosan films for active packaging *Ind. Crops Prod.* **94** 800–11
[42] Sipponen M H, Lange H, Crestini C, Henn A and Österberg M 2019 Lignin for Nano- and Microscaled Carrier Systems: Applications, Trends, and Challenges *ChemSusChem* **12** 2038
[43] Zikeli F, Vinciguerra V, Sennato S, Scarascia Mugnozza G and Romagnoli M 2020 Preparation of Lignin Nanoparticles with Entrapped Essential Oil as a Bio-Based Biocide Delivery System *ACS Omega* **5** 358–68
[44] Vettraino A M, Zikeli F, Scarascia Mugnozza G, Vinciguerra V, Tabet D and Romagnoli M 2022 Lignin nanoparticles containing essential oils for controlling Phytophthora cactorum diseases *For. Pathol.* **52** e12739

Methodology for the investigation of surfactant dissolution in water

R Ferraro[1] and S Caserta[1]
[1]DICMaPI, Università di Napoli Federico II, P.le V. Tecchio 80, 80125 Napoli, Italy

Abstract. Concentrated aqueous solutions of surfactants pastes are widely used in industrial productions. The dissolution phenomena of these fluids in water is difficult to optimize, given the thermodynamic complexity of the system, that adds further resistance to the mass transport efficiency. One of the most common anionic surfactants is Sodium Lauryl Ether Sulphate (SLE_3S). Depending on the SLE_3S-water composition, surfactant nano-molecules can assume different morphologies, such as micellar phases (<30%wt of SLE_3S), hexagonal liquid crystals (>30%wt), cubic structures (~60%wt), or lamellar liquid crystals (>60%wt), which are characterized by nanoscale ordered structures. Different phases show strongly different rheological behaviour. All phases are non-Newtonian, with a shear-thinning viscosity as function of imposed shear rate. A novel methodology is here proposed to investigate the dynamic evolution of the phase changes during surfactant dissolution using Time-lapse microscopy. Samples are observed trough crossed polarizers. Differences in nano-morphology, due to molecules rearrangement during phase changes, can be easily monitored by measuring light intensity profiles.

1 Introduction

Surfactants are amphiphilic molecules characterized by a hydrophobic tail and a hydrophilic head, that are fundamental for an extremely wide range of applications, including home- and beauty- care, everyday use goods [1-3], food [4], and pharmaceutical industry [5]. Due to their complex molecular structure, as the surfactant concentration in a water solution increases, the system undergoes a series of phase changes [6]. Despite the relevance of the topic, the phenomenon is still far to be fully understood.

The main goal of this work is to fill this gap in the knowledge, developing a methodology for the systematic characterization of the dissolution phenomenon in static conditions. As case study we referred to one of the most used anionic surfactants: Sodium Lauryl Ether Sulphate (SLE_3S). SLE_3S is one of the main components in shampoos, detergents, toothpastes, and creams, used for its emulsifying and foaming properties and for its reduced production costs. Thus, establishing the behaviour of this commercial system is of outstanding relevance for several industrial fields. SLE_3S is a sodium salt produced through the ethoxylation of the Sodium Lauryl Sulphate (SLS). According to typical behaviour of surfactant systems, SLE_3S rearrange in different phases, depending on concentration in water. In detail, at room temperature (25°C), phase changes are defined [7, 8] in the following concentration ranges: micellar (referred as L_1) in the range 0.024-28%wt, hexagonal (H) 31-56%wt, cubic (V_1) ~58-62%wt and lamellar (L_α) 62-72%wt. Molecular rearrangement on the nano-scale lead different optical behaviour of the sample. In particular, when observed trough crossed polarizers different phases appear with different luminosity, due to differences in birefringence. A scheme of phase diagram at room temperature of this system is reported in Figure 1, where each phase (L_1, H, V and L_α), separated by dotted lines, is identified by a schematization of the molecular structure, coupled with an optical image, acquired through crossed polarizers. From left to right: Micellar structure (L_1) is characterized by a dark region; Hexagonal phase (H) marked by packed rods, where higher birefringence induces higher luminosity; dark isotropic cubic phase (V) where molecules are packed in cubic arrays. Finally, a virtually infinite bi-dimensional multi-layer structure characterized by alternating aqueous and surfactant layers where hydrophilic and hydrophobic groups are aligned, this phase is defined as lamellar phase (L_a). Multi lamellar domains can also fold and form vesicles [9], that and under crossed polarized light presents typical Maltese cross patterns.

In our work, Time-Lapse polarized light microscopy is combined with rheological characterization to study the phase transitions taking place during the dissolution in water. Concerning rheology, concentrated surfactant-water solutions, usually defined as surfactant pastes, show a gel-like behaviour. Heavily different rheological behaviour of the fluid is observed for different water-surfactant concentrations, with non-monotonic variations in viscosity of several orders of magnitude.

2 Materials and Methods

2.1 Sample preparation

SLE_3S was obtained from Procter and Gamble. The paste contains 72% weight fraction of surfactant in water, as determined by vacuum drying. Sample at different concentration were prepared using the amount of bi-distilled water required to dilute the surfactant raw paste. Each diluted sample was mixed with a spatula and centrifugated (4000 rpm for 40 min) several times up to obtain complete degassing. Samples were then stored at room temperature overnight to reach the equilibrium and checked by visual inspection through cross-polarizers, obtaining images analogous to the examples reported in Figure 1.

2.2 Rheology setup

Rheological measurements were performed at room temperature using a stress-controlled rheometer (Anton Paar Physica MCR 301 Instruments) equipped with a cone-plate measuring system (CP25-1/S-SN72510). Pre-shear was run to homogenise sample structure after loading. Flow curves were run varying shear rate in the range 0.01–200 s^{-1}.

2.3 Time-Lapse workstation

SLE_3S phases evolution was followed by automated Time-Lapse Microscopy [10] based on an inverted microscope (Zeiss Axiovert 200, Carl Zeiss, Jena Germany). Several independent fields of view were acquired by a high-resolution high-sensitivity monochromatic CCD video camera (Hamamatsu Orca AG, Japan) using a 10x objective and two crossed polarizers to visualize the SLE_3S internal nanostructure. The workstation was also equipped with motorized stage and focus (Marzhauser) controlled with a home-made LabVIEW code for automated mosaic scanning (e.g., 5x4 images, corresponding to a scanned area of about 4x2.5 mm) of large samples. A workstation scheme is reported in Figure 2.

2.4 Microchannel preparation

SLE_3S dissolution in water was studied in a linear geometry configuration [11] (schematized in Figure 2). A microchannel height 200 μm and wide 5 mm, obtained by modifying a commercial bottomless channel (Ibidi) was used. The resulting device is characterized by two openings: the first is used to introduce the surfactant at 72 %wt and the other, a small chamber realized by using an O-ring, hosts a bi-distilled water reservoir. Images, at room temperature, were acquired at regular intervals (10 min), up to the complete dissolution of concentration in the entire chamber.

2.5 Image Data Analysis

The dissolution process of SLE_3S in water was quantified by measuring the light intensity profile by averaging the grey level along y-direction at each x position (where the interface corresponds to x=0 μm). The mean grey levels (MGL), normalized respect to its value in water (MGL_w), was plotted as function of the sample length (x, μm), in order to identify and distinguish each phase.

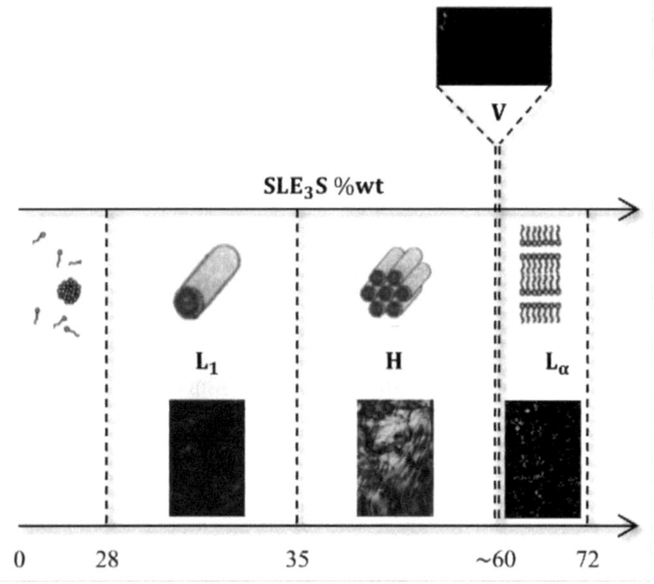

Figure 1. Scheme of phase diagram of SLE_3S-water system: optical images of each phase - micellar (L_1), hexagonal (H) and cubic (L_α) – are reported for sample of ~130 μm thickness.

IOP Conf. Series: Materials Science and Engineering 1265 (2022) 012010 doi:10.1088/1757-899X/1265/1/012010

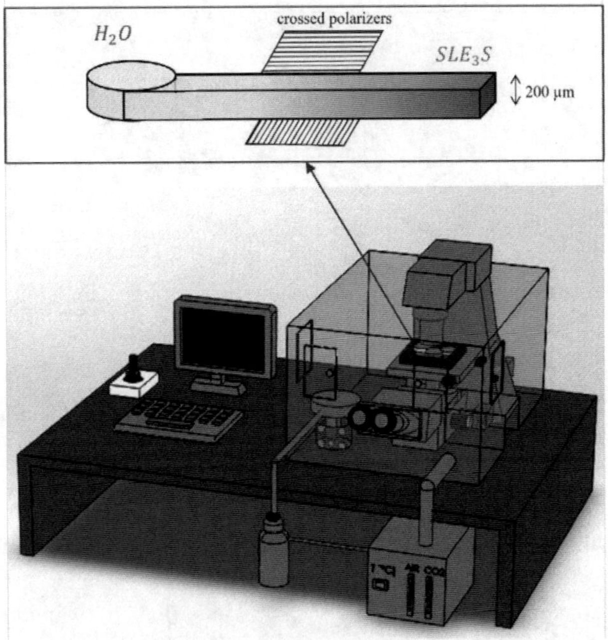

Figure 2. Scheme of Time-Lapse workstation and microchannel configuration.

3 Results

A wide variety of phases can be realized by mixing SLE_3S and bi-distilled water, depending on the ratio of the components. Recognition of different microstructure, from micellar to lamellar, can be achieved by microscopy (optical and dissolution experiments) and rheology. In the following section the experimental results obtained applying our experimental method are reported.

3.1 Rheological characterization

In Figure 3, steady viscosity data (σ, Pa) as a function of the shear rate ($\dot{\gamma}$, 1/s) are shown for the three concentrations here in exam (35, 60 and 72%wt), corresponding to different phases (H, V_I and L_α, respectively), at room temperature. In agreement with previous measurements reported in the literature [8], a non-Newtonian behaviour, in particular shear–thinning, is observed. It is worth mentioning the rheological behaviour is non-monotonic increasing SLE_3S concentration. In particular in the concentration range where cubic phase is present (~60%) viscosity is several orders of magnitude higher respect to the other two phases (H and L_α).

IOP Conf. Series: Materials Science and Engineering 1265 (2022) 012010 doi:10.1088/1757-899X/1265/1/012010

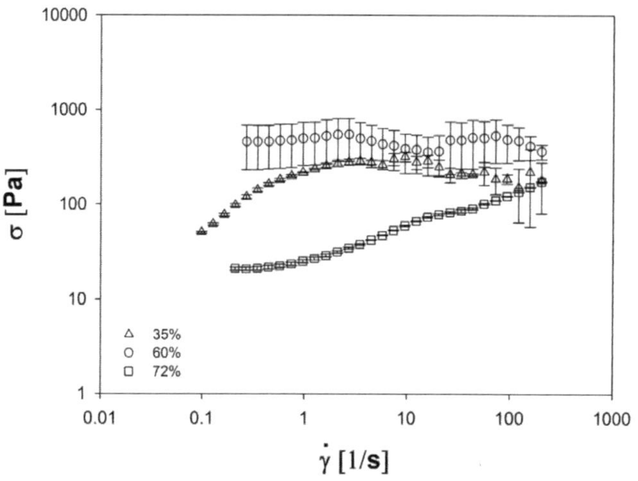

Figure 3. Flow curves for various SLE_3S concentrations (35%wt – triangle symbols, 60%wt – circle symbols and 72%wt – square symbols) at room temperature.

3.2 Dissolution

After adding water in the microchannel reservoir, diffusive flow of water in the channel induces progressive dissolution of the surfactant paste sample. As time goes on, sample evolves in different phases, due to change in water concentration. Consequently, different regions of the same sample have different birefringence, visible as vertical strips in the mosaic image. In Figure 4, different images acquired at different times during the dissolution experiment are reported ($t_{0,1,2,3}$). In detail, in the image acquired at the beginning of dissolution experiment (t_0) the entire sample show a uniform morphology and light intensity, as consequence of the uniform concentration of the paste in lamellar phase (L_a). The structure is characterized by Maltese cross patterns, typical of multilamellar vesicles [9]. During the dissolution ($t_{1,2,3}$), the diffusion driven onset of a water concentration profile within the rectangular channel induces the formation of different phases along the sample. From left to right, a dark region corresponds to the micellar phase (L_1) domain. A second band, where higher birefringence induces higher luminosity, represents the surfactant in the hexagonal phase (H) that borders with a third region that, according to the equilibrium phase diagram [12], can be associated to the isotropic cubic phase (V). It is interesting to note that the third band (H phase) shows a not uniform brightness and tends to become increasingly darker (isotropic) as it approaches the edge of the micellar phase. The last band on the right edge of the channel is the lamellar phase, being invaded by the waterfront. As time goes on each band moves from the water reservoir to the right with a different diffusive coefficients for each phase [8].

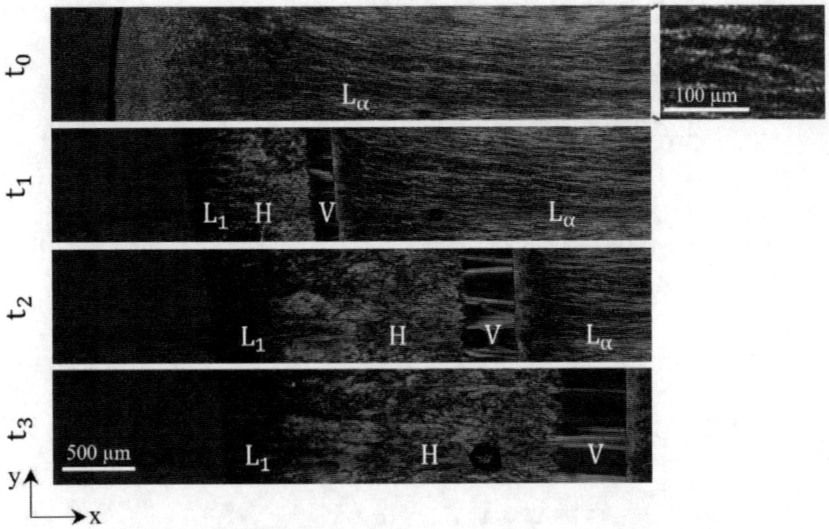

Figure 4. PLM images of dissolving SLE_3S paste at four different time ($t_{0,1,2,3}$). At time t_0, an image of sample in lamellar phase before adding water. Three frames, $t_{1,2,3}$, recorded after the beginning of dissolution, corresponding to 15, 50 and 85 min, respectively, after the initial contact of water with surfactant paste (t_0) are reported. In each frame all the phases - micellar (L_1), hexagonal (H), cubic (V_1) and lamellar (L_α) – can be identified.

In order to obtain more quantitative information about the brightness levels of the image reported in Figure 4, the light intensity profile along x direction was calculated by averaging the grey level along y-direction. Normalized light intensity profiles (MGL/MGL_w) are reported in Figure 5 as function of the sample length (x, µm) for each time of dissolution experiment ($t_{0,1,2,3}$). Each phase corresponds to a specific section in the plot, highlighted by dotted lines. The intensity profile allows to identify and characterize all the above-mentioned bands. Specifically, micellar (L_1), cubic (V), and lamellar (L_a) phases are characterized by a constant and low value of intensity, while hexagonal (H) phase shows value scattered throughout the channel. Analogously, the passage from the cubic to the hexagonal phase is not sharp but occurs over a finite x range. More detailed analysis of dissolution phenomena is available in a previous paper where the same technique here presented is coupled also with Raman spectroscopy to quantify concentration profiles [11].

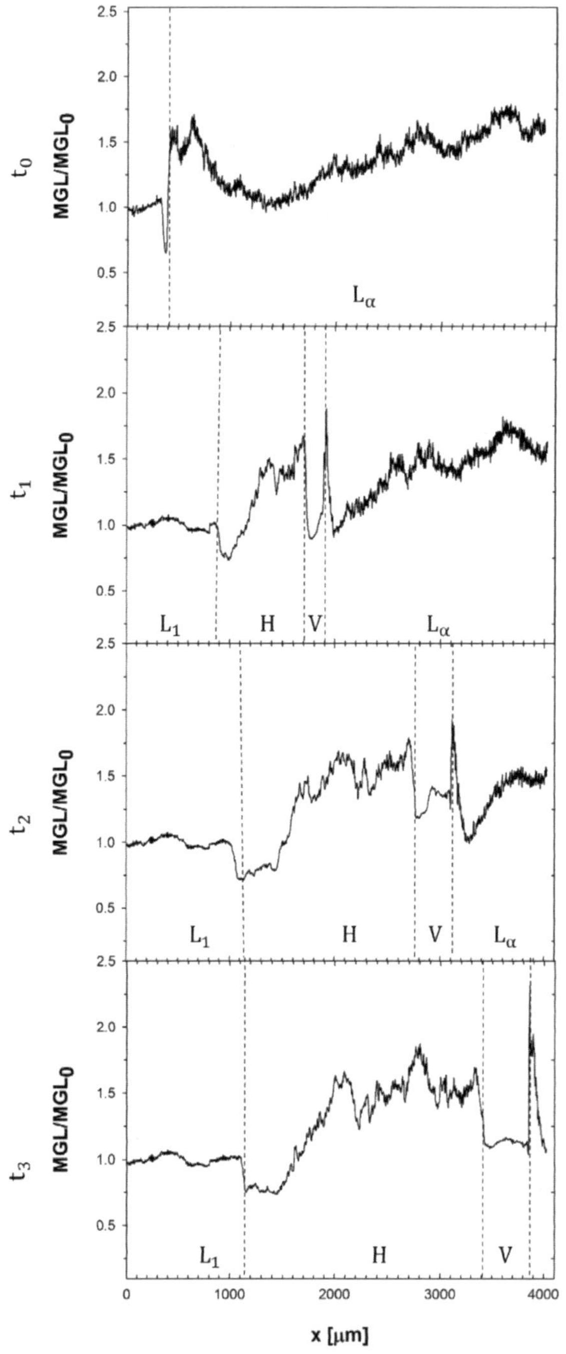

Figure 5. Time evolution of the intensity band. In a frame recorded at the beginning of dissolution (t_0) only the lamellar structure is evident, while after the dissolution ($t_{1,2,3}$) all the phases (micellar L_1, hexagonal H, cubic V_1 and lamellar L_α) have developed. (t_1=15 min, t_2=50 min and t_3=85 min, measured from t_0, corresponding to the initial contact of water with surfactant paste).

Conclusions

In this work, a methodology for the investigation of surfactant dissolution in water by combining two techniques (polarized light microscopy and rheology) is proposed. The dissolution of SLE_3S in water is a complex process because several phase transitions occur as surfactant concentration decreases with time, involving significant physical-chemical changes.

The different phase changes happening during the dissolution process do not occur suddenly but are associated to an intermediate (inter-phase) molecular reorganization [11], that is visible by the evolution of light intensity profile of images acquired through cross-polarizers. In fact, polarized light microscopy provide details on nanostructure of the sample, linked to anisotropies in the optical properties.

References

[1] Baruah A, Pathak A K and Ojha K 2015 Phase behaviour and thermodynamic properties of lamellar liquid crystal developed for viscoelastic surfactant based fracturing fluid *Chemical Engineering Science* **131** 146-54

[2] Hargreaves A E 2007 Chemical formulation: an overview of surfactant based chemical preparations used in everyday life

[3] Hull K L, Sayed M and Al-Muntasheri G A 2016 Recent advances in viscoelastic surfactants for improved production from hydrocarbon reservoirs *SPE Journal* **21** 1340-57

[4] Drakontis C E and Amin S 2020 Biosurfactants: Formulations, properties, and applications *Current Opinion in Colloid & Interface Science* **48** 77-90

[5] Garti N, Libster D and Aserin A 2012 Lipid polymorphism in lyotropic liquid crystals for triggered release of bioactives *Food & function* **3** 700-13

[6] Richtering W 2001 Rheology and shear induced structures in surfactant solutions *Current opinion in colloid & interface science* **6** 446-50

[7] Li H, Dang L, Yang S, Li J and Wei H 2016 The study of phase behavior and rheological properties of lyotropic liquid crystals in the LAS/AES/H2O system *Colloids and Surfaces A: Physicochemical and Engineering Aspects* **495** 221-8

[8] Castaldo R I, Pasquino R, Villone M M, Caserta S, Gu C, Grizzuti N, Guido S, Maffettone P L and Guida V 2019 Dissolution of concentrated surfactant solutions: from microscopy imaging to rheological measurements through numerical simulations *Soft matter* **15** 8352-60

[9] Pommella A, Caserta S, Guida V and Guido S 2012 Shear-Induced Deformation of Surfactant Multilamellar Vesicles *Physical Review Letters* **108** 138301

[10] Vuoso D C, D'Angelo S, Ferraro R, Caserta S, Guido S, Cammarota M, Porcelli M and Cacciapuoti G 2020 Annurca apple polyphenol extract promotes mesenchymal-to-epithelial transition and inhibits migration in triple-negative breast cancer cells through ROS/JNK signaling *Scientific reports* **10** 1-17

[11] Capaccio A, Caserta S, Guido S, Rusciano G and Sasso A 2020 Dissolution of a surfactant-water lamellar phase investigated by combining time-lapse polarized light microscopy and confocal Raman spectroscopy *Journal of colloid and interface science* **561** 136-46

[12] Poulos A S, Jones C S and Cabral J T 2017 Dissolution of anionic surfactant mesophases *Soft matter* **13** 5332-40

Comparison of different correlative AFM-SEM workflows on calcite moonmilk

S Dinarelli[1], F Mura[1,2], C Mancini[1], G La Penna[1], T Rinaldi[2,3] and M Rossi[1,2]

[1] Department of Basic and Applied Science, Sapienza University of Rome, Rome, Italy
[2] CNIS - Center on Nanotechnology Applied to the Engineering of Sapienza, Sapienza University of Rome, Rome, Italy
[3] Departiment of Biology and Biotechnology, Sapienza University of Rome, Rome, Italy

E-mail: francesco.mura@uniroma1.it

Abstract. In recent years, high resolution microscopy techniques are evolving toward a fast combination of different microscopies and spectroscopies, generally labelled under the title of correlative microscopy, each capable to provide unique information and a more comprehensive characterization of the sample under analysis. Among them stands out the Correlative Probe to Electron Microscopy (CPEM), where Scanning Electron Microscopy and Scanning Probe Microscopy are combined. This kind of technique is relatively new, and its range of capabilities is still not fully explored. In this paper, a demonstration of different CPEM workflows to characterize the moonmilk, a particular type of nanostructured calcium carbonate, extracted from ancient tombs of the Etruscan Necropolis of Tarquinia, is provided. Besides, the advantages of an innovative AFM-in-SEM setup, even respect to the standard standalone AFM measurement, are presented, showing how the analysis of the moonmilk nano-fibers, a rather challenging sample to be analysed with probe microscopies, is simplified and with less risk of artefacts or contamination of the AFM probe.

1. Introduction

During the last decade, the so-called correlative microscopy has gained increasing relevance in both research and industry, providing not only a strong impulse for the creation of new specific instrumentation, but also towards the development of rapid and effective analysis protocols for its application [1]. The key component of this experimental approach is the precise and fast localization of the very same features among the different microscopic and spectroscopic techniques in use. This is not always straightforward and, essentially, it can be performed following two different approaches: (a) to implement "in-situ" several techniques and perform the measurement at the same time or sequentially without moving the sample or (b) implement a strategy to localize the same region of the sample while physically moving it among the different instrumentations, for example by means of localizing grids or markers. The first and most known example of correlative microscopy is the CLEM – Correlative Light to Electron Microscopy – where the high-resolution imaging of the electron microscope is correlated with a fluorescence optical analysis, thus granting the possibility to highlight information regarding the internal functions of cells and tissues [2]. However, not only light and electron microscopies can be combined, but by adopting specific protocols, a great number of combinations with different characterizing techniques, such as Raman or Atomic Force Microscope (AFM) can be done [3,4]. All the available configurations make the correlative microscopy a very powerful tool, capable to improve the knowledge of complex phenomena that can't be achieved by the standalone microscopies, while, at the same time, it avoids dispersion of data due to different sample preparations and treatments [1]. Moreover, this approach is not limited only to a normal bidimensional analysis, but it can be arranged also to perform a complete 3D reconstruction of a sample, by combining, for example, X-Ray tomography with Focused Ion Beam and Electron Tomography [5]. In this paper we take advantage of the capabilities of a recently developed correlative microscopy that is called CPEM – Correlative Probe to Electron Microscopy –, that consists in the combination of Scanning Electron Microscopy (SEM) with a Scanning Probe Technique [6,7]. In our experimental configuration, we have an AFM operating in tapping mode placed inside the chamber room of the electron microscope. In this way, SEM imaging is used to quickly navigate the sample and localize the Region of Interest (ROI) in which the AFM measurement will be performed. Besides, during the AFM scan, the electron beam is maintained on and focused a few nanometers away from the tip, generating a very slow SEM image in parallel with AFM measurement. This gap between the tip and the electron beam is necessary to avoid the disturbance of the tip itself (shadowing effect) [8]. Then, from the overlapping of these two images, it is possible to obtain a final CPEM image that condensate on the very same area, all the information typical of the SEM imaging with the real height values measured with the AFM. However, it is worth mentioning that the AFM can be used not only to measure the height, but also others specific physical properties such as local elasticity or local conductivity [9]. In this paper, we used Raman spectroscopy, AFM and SEM to characterize two different samples of moonmilk, a particular type of nanostructured calcium carbonate, sampled from ancient tombs of the Etruscan Necropolis of Tarquinia [10,11]. Moreover, the capabilities of CPEM will be discussed by comparing the results obtained using two different and physically separate SEM and AFM and an innovative AFM-in-SEM system, the Litescope from Nenovision [12]. The rationale of this work is to demonstrate the usefulness of the CPEM technique onto a very challenging samples from the AFM point of view, since moonmilk consists in relatively variable sized objects, deposited onto a rough surface and located at different height planes. This kind of samples can rapidly deteriorate the AFM tip, compromising the whole measurements. In this panorama, it is crucial to avoid performing "blind" AFM analysis, i.e., looking at random region of the sample trying to find the object of interest, but instead, the correlative approach could be an enormous advantage, providing a preliminary zoom in the region of interest with the SEM and directly checking in live mode the position of the AFM tip and start the AFM measurements at the precise desired location.

2. Materials and Methods

2.1. Environmental parameters for the sample collection and SEM\AFM preparation

Moonmilk deposits, about 0,5 gr., were collected from two ancient Etruscan tombs: Tomba degli Scudi and Tomba Maggi located in the Monterozzi Necropolis [13], using sterile scalpels and kept in 10 ml sterile tubes. The relative humidity in the tombs typically varied in the range 95-98% and the temperature between 16-18°C. A little portion of powder containing the sample has been deposited on an aluminium stub, previously cleaned with acetone, partly covered by bi-adhesive tape. The deposition has been performed in such a way that a part of the sample stitch to the tape and a part remains directly on the metal. The obtained stubs have been used throughout all the experiments performed in this paper. Since the two tombs were carved in two different rocks, the samples will be defined according to the rock's substrate name, i.e., Macco (fossiliferous calcarenite) and Sabbione (hybrid sandstone) that belongs to the Tomba Maggi and Tomba degli Scudi, respectively [11,14].

2.2. Instrumentation and setups

2.2.1. Scanning Electron Microscope and Focus Ion Beam

The orientation marker for the ex-situ CPEM workflow has been produced with a Cobra Focused Ion Beam (from Physics D'Orsay), mounted on an Auriga 405 FE-SEM (from Zeiss). The operating condition of the electron microscope for this analysis are 3 kV for the acquisition voltage, 5.5 mm for the working distance (i.e. the distance between the electron gun and the sample), while high vacuum has been kept at a pressure of 10^{-5}-10^{-6} mbar.

2.2.2. *Stand-alone AFM*

The AFM stand-alone measurements were performed with an AFM ICON (from Bruker) in two modalities: (a) contact mode while equipped with triangular-shaped silicon nitride cantilever (model DNP), with nominal elastic constant of 0.06 N/m and (b) tapping mode while equipped with rectangular-shaped silicon nitride cantilever (model RTESP) with nominal resonant frequency of 300 kHz and elastic constant of 40 N/m. Both cantilevers ended with a silicon nitride tip with nominal radius of 10 nm. The maximum vertical force used has been set lower than 1 nN to avoid damages to the tip or the sample.

2.2.3. *AFM-in-SEM*

To acquire in situ CPEM images, a combined system made of SEM VEGA (from TESCAN) and an AFM LiteScope (from Nenovision) has been used. The SEM is equipped with a tungsten's filament while the Litescope module is allocated directly on the sample holder inside the SEM chamber by means of 4 screws and a proprietary adaptor, the complete setup can be seen in [12]. AFM measurements have been performed in tapping mode with a self-sensing and self-actuating tuning-forkbased tip (model Akiyama) with nominal resonant frequency of 45 kHz and nominal elastic constant of 5 N/m. To obtain the SEM micrograph for the final CPEM image, the electron beam is taken active and focused on a position few nanometres away from the tip, to avoid the creation of any kind distortion due the tip itself, the so-called shadowing effect [8]. Throughout all the characterizations the chamber in maintained in high vacuum at a pressure of 10^{-4} mbar, to grant the possibility of obtaining SEM images of good quality. The acquisition voltage was set to 4-6 kV in such a way to not damage the sample itself or create artifacts in the acquired images, while the working distance was set to around 10mm in order to maximize the signal-to-noise ratio.

2.2.4. *Micro-Raman*

Raman spectra were collected using a confocal inVia™ Raman Spectrometer (from Renishaw), with 250 mm focal length. The specimens were analyzed at room temperature in the 145-1800 cm^{-1} spectral range. The signal is dispersed by a holographic grating of 1800 l/mm and collected by a Peltier-cooled CCD detector. The excitation line at 532.1 nm was produced by a Nd:YAG continuous-wave diode-pumped solid state laser (from Renishaw) and focused on the sample through a short distance working objective N PLAN 50x, with NA = 0.75 (from Leica Microsystems). The laser power impinging on the samples was set to 25 mW with an exposition time of 0.4s and 15-20 accumulations. For each sample five spectra have been collected. The peak positions were automatically calibrated using an internal reference of a Ne-Ar light available in the spectrometer. The software WiRETM 4.4 has been used to normalize the spectra, perform fitting procedures, and obtain peaks position. Normalized intensities of each band are obtained considering the ratio between their height and the intensity of the most intense band in the spectrum.

3. Results and Discussions

The Etruscan Tombs in the Monterozzi Necropolis of Tarquinia, carved in calcarenite and hybrid sandstone rocks, with constant temperature and humidity, show the presence of nanometric-scale crystals of calcium carbonate in the form of calcite, forming a white patina covering the walls; this speleothem is called moonmilk and, despite its formation has a biogenic origin, the physical mechanism promoting the formation of the nanometric fibres is unknown [14-16]. In previous studies, we analysed a collection of samples from 13 tombs of the archaeological area, and we observed that the rock substrate has a strong influence on the final structures of the moonmilk nanorods, showing, for example, that in the Tomba degli Scudi, excavated in a hybrid sandstone vein, the moonmilk is made of mostly of single and multiples smooth nanorods [11].

3.1. Raman characterization

The Raman spectroscopy has been performed on the moonmilk samples collected from two Etruscan tombs having a different rock substrate, as further confirmation of the analysis reported in a previous paper [14]. This is a further proof that the shape and size of the nanofibers doesn't change its crystalline structure. In fact, the moonmilk is a biogenic form of the calcite, which is characterized by a rhombohedral geometry, precisely a D_{3d}^6 point symmetry [17], and, according to the group theory of vibrational spectra [18], the Raman spectrum of these minerals is expected to have five fundamental phonons, listed in Table 1.

Table 1. Description of the five fundamental phonons of calcite [19].

Phonon symbol	Wavenumber [cm^{-1}]	Description
$\mathbf{E_g(T)}$	154	*External vibration* due to the interaction between CO_3^{2-} and Ca^{2+}: translations of the CO_3^{2-} group in the primitive cell around the axes normal to the C_3 axis.
$\mathbf{E_g(L)}$	281	*External vibration* due to the interaction between CO_3^{2-} and Ca^{2+}: liberations of the CO_3^{2-} group in the primitive cell around the axes normal to the C_3 axis.
$\mathbf{E_g}$	711	*Internal vibrations* of the CO_3^{2-} group.
$\mathbf{A_{1g}}$	1087	
$\mathbf{E_g}$	1438	
$\mathbf{E_g+A_{1g}}$	1742	A combination band.

The obtained spectra, showed in fig. 1, confirm the presence of the calcite, where the four peaks positions expected in the range 155-1086 cm^{-1} are perfectly overlapped with the reference spectrum, also respecting the intensity ratio among the peaks.

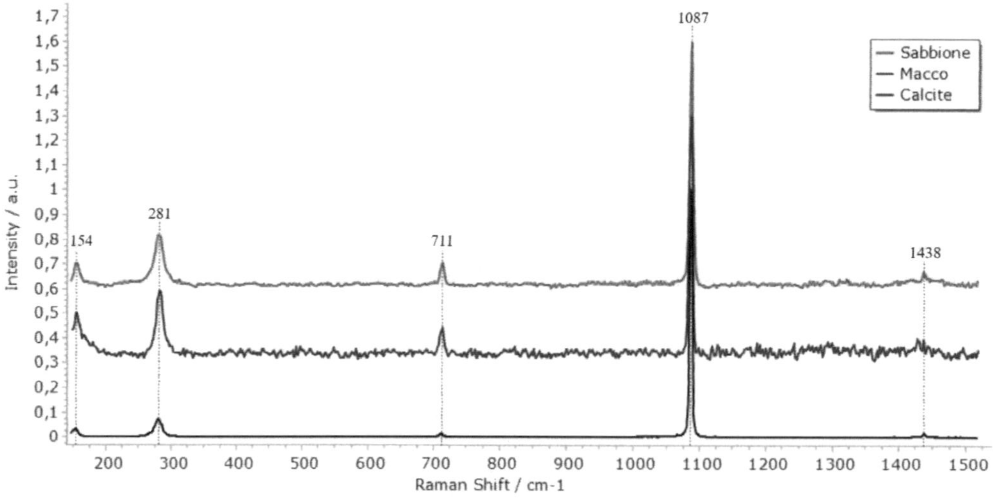

Figure 1. Comparison (in order from top to bottom) between the spectra acquired on Sabbione (red line), Macco (blu line) and reference calcite spectrum from the Renishaw Database (black line) – color online. The spectra are normalized with respect to the peak at 1087 cm^{-1} and stacked with a vertical shift of 0.3 a.u..

3.2. *CPEM characterization*

The workflow of a typical CPEM measurement operated through a reference marker, as performed in this work, is reported in figure 2.

Figure 2. Subsequent steps to perform a CPEM image on a moonmilk sample using a FIB reference marker. (a) and (b) progressive zoom with the FESEM of the very same region subsequently imaged with the AFM; (c) higher magnification of a single moonmilk nanofiber and (d) corresponding AFM image on the structure of interest acquired with the AFM-in-SEM configuration.

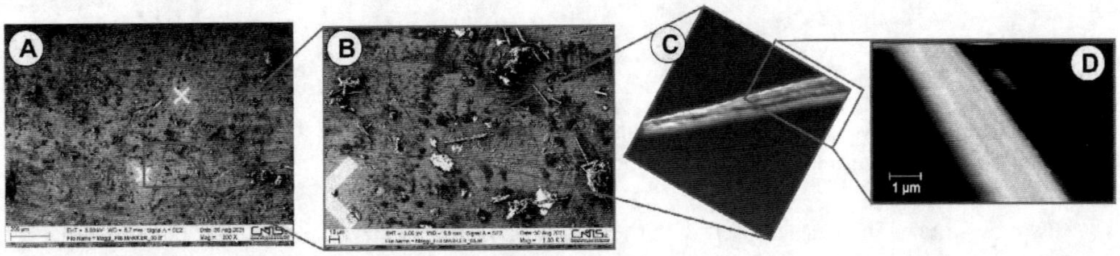

In figure 2(a) and 2(b) are reported two FESEM images of the region of interest taken at different magnifications, where the FIB markers, the two crosses, are clearly visible. The same sample has been transferred to the SEM from TESCAN, to perform the AFM-in-SEM characterizations. In figure 2(c), it's reported a higher magnification of a specific single moonmilk nanofiber, whose image has been rotated respect to its original orientation inside the TESCAN microscope chamber room to give a clear connection with the previous two images of the figure. This area has been later scanned and observed as an AFM image (fig. 2(d)). However, the initial step of the creation of the FIB marker can be avoided, as well as their search through the optical camera of the AFM, by performing an AFM-in-SEM approach. In our case, the Litescope module has been mounted directly inside the chamber room of the SEM, and besides being a timesaving approach to the measurement, it allows to take advantage of the capabilities of the SEM to have a fast scan of the sample and then to have a fast identification of the region of interest. In this way, the AFM image can be acquired with absolute certainty of the positioning of the tip and even with the selection of the right image size. An example of this "in-situ" CPEM approach is shown in fig. 3, where an area of 20 μm^2 containing some moonmilk nanofibers has been analysed.

Figure 3. Composition of a CPEM image: (a) SEM and (b) AFM raw images acquired simultaneously. (c) 2D and (d) 3D representation of the CPEM image.

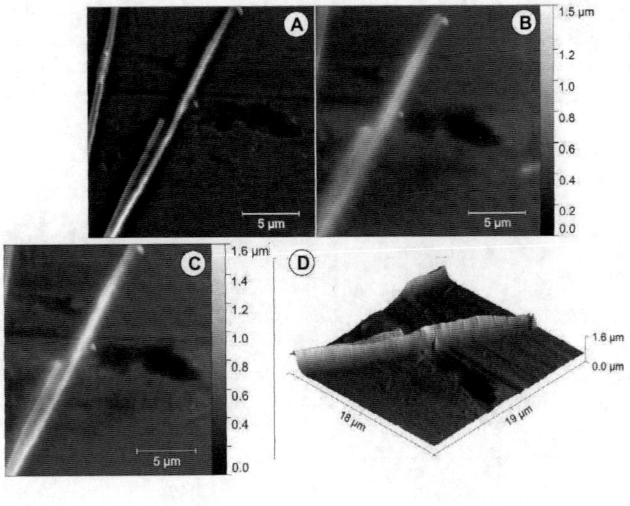

Looking at the fig. 3(a) and 3(b), it's easy to notice that the SEM and AFM images are slightly shifted. This shifting is intrinsic of the technique, since, to make possible the simultaneous acquisition of a SEM image during the AFM scanning, the electron beam is maintained at a fixed distance from the tip, to avoid the shadowing effect. However, this issue is easy to overcome by superimposing the two acquisitions, as shown in fig. 3(c) and 3(d), where the final CPEM image in 2D and 3D view are, respectively, reported. By the comparison of all the pictures exposed in fig. 3, it's easy to notice that CPEM is not only an addition of the qualitative height measurement of the AFM, but an enhanced image able to embed the best from both instruments, resulting into a more complete characterization of the sample.

3.3. Standalone AFM analysis

For comparison, we have analysed the two samples adopting a standard stand-alone AFM approach, where some acquired areas are reported in figure 4.

Figure 4. AFM images acquired with the standalone AFM in air: (a) tapping mode image of Sabbione, (b) contact mode and (c) tapping mode images acquired on the Macco's sample.

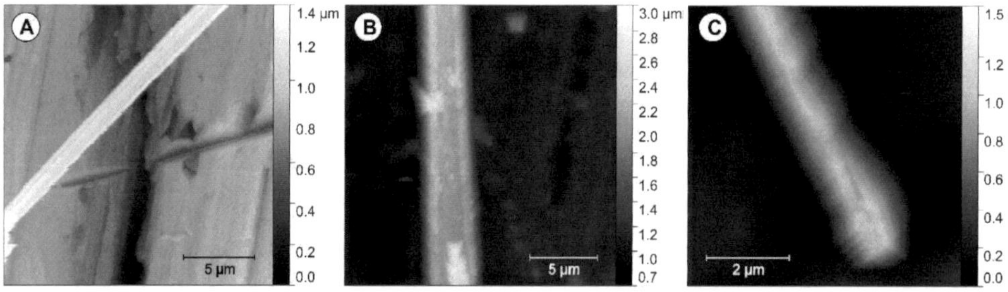

The quality of the images is almost the same as presented in the previous chapter, but with two crucial differences: (a) to obtain an image of a single structure, several areas need to be scanned to locate a structure of interest; (b) once a structure is located, you have no control on the surrounding, and could happen that this nanofiber can be stacked on the top of another, overcoming the z-piezo vertical range. These effects can lead to a fast deterioration of the AFM tip, like collecting dirt from the substrate or even collecting entire pieces of the nanofibers. One example is illustrated in fig. 4(a), where an AFM image of the Sabbione sample in tapping mode is reported. The final image is very neat and sharp, but a small defect can be noted in the bottom left, maybe due to the presence of another superimposed structure. Other examples of tip contamination are shown in figure 4(b) and 6(c), where two different areas have been acquired onto the Macco sample, respectively, in contact and tapping mode. In both cases, the presence of some contaminations on the tip influences the achievable resolution, and it should be noted that the images here reported are the best ones obtained in several days of measurements with the stand-alone AFM, another important evidence of the very challenging nature of these class of samples in Atomic Force Microscopy.

4. Conclusions

In this paper we have reported two different workflows for the CPEM characterization applied to an AFM challenging sample like the moonmilk, illustrating the advantages that arise from a direct measurement through an AFM-in-SEM device. These nanofibers have been extracted from two Etruscan tombs, having a different rock substrates underneath, the Macco for Tomba Maggi and Sabbione for Tomba degli Scudi, which have a great influence on the final shape and size of the moonmilk. The

Raman analysis confirms that they are made of calcite, without showing any kind of variation due to their different morphology. Besides, the capabilities of our AFM-in-SEM system of approaching the CPEM in an easiest and fastest way, as well as ensuring additional information, respect using a traditional AFM or a FIB marker workflow, have been demonstrated. Another important aspect of an in-situ CPEM is the possibility to reduce dramatically the presence of artifacts and contaminations of the tip, that are unavoidable in the standard "blind" AFM analysis. For these reasons, further studies will be devoted to the high-resolution systematic characterization of the two samples, to provide quantitative data on the structural differences, like the roughness, of the different type of moonmilk.

Acknowledgments

This research has been realized with the financial support of the projects CHALLENGES - Real time nano characterization related technologies, granted by European Commission (call H2020-NMBP-TO-IND-2019) and ATOM - Advanced Tomography and Microscopies, granted by Lazio Region (call "Open infrastructures for research").

References

[1] Parlanti P and Cappello V 2022 Microscopes, tools, probes, and protocols: A guide in the route of correlative microscopy for biomedical investigation *Micron* **152** 103182.

[2] Giepmans BNG 2008 Bridging Fluorescence microscopy and electron microscopy *Histochem. Cel. Biol.* **130** pp. 211-217.

[3] Schmidt R, Nachtnebel M, Dienstleder M, Mertschnigg S, Schroettner H, Zankel A, Poteser M, Hutter H, Eppel W and Fitzek H 2021 Correlative SEM-Raman microscopy to reveal nanoplastics in complex environments *Micron* **144** 103034.

[4] Jadavi S, Bianchini P, Cavalleri O, Dante S, Canale C and Diaspro A 2021 Correlative nanoscopy: A multimodal approach to molecular resolution *Microsc. Res. Tec.* **84** pp. 2472-2482.

[5] Ma L, Dowey P, Rutter E, Taylor K and Lee P 2019 A novel upscaling procedure for characterising heterogeneous shale porosity from nanometer-to millimetre-scale in 3D *Energy* **181** pp. 1285-1297.

[6] Holz M, Reuter C, Reum A, Ahmad A, Hofmann M, Ivanov T, Mechold and Rangelow IW 2019 Atomic force microscope integrated into a scanning electron microscope for fabrication and metrology at the nanometer scale *Proc. SPIE Photomask Technology 2019* Vol. 11148 p. 111481F.

[7] Rangelow IW, Kaestner M, Ivanov T, Ahmad A, Lenk S, Lenk C, Guliyev E, Reum A, Hofmann M, Reuter C and Holz M 2018 Atomic force microscope integrated with a scanning electron microscope for correlative nanofabrication and microscopy *Journal of Vacuum Science & Technology B* **36(6)** 06J102.

[8] Fukushima K, Kawai S, Saya D and Kawakatsu H 2002 Measurement of mechanical properties of three-dimensional nanometric objects by an atomic force microscope incorporated in a scanning electron microscope *Rev. Sci. Instrum.* **73(7)** pp. 2647 – 2650.

[9] Novotna V, Horak J, Konecny M, Hegrova V, Novotny O, Novacek Z and Neuman J 2020 AFM-in-SEM as a Tool for Comprehensive Sample Surface Analysis *Microscopy Today* **28(3)** pp. 38-46.

[10] Tomassetti MC, Cirigliano A, Arrighi C, Negri R, Mura F, Maneschi ML, Gentili MD, Stirpe M, Mazzoni C and Rinaldi T. 2017 A role for microbial selection in frescoes' deterioration in Tomba degli Scudi in Tarquinia, Italy *Scientific Reports* **7** 6027.

[11] Mura F, Cirigliano A, Maras D and Rinaldi T, 2021 Analysis of moonmilk nanofibers in the etruscan tombs of Tarquinia, *AIP Conference Proceedings* **2416** 020014.

[12] https://www.nenovision.com/

[13] Cecchini A, Adamo F, Buranelli F and Cataldi M 2012 *Le tombe dipinte di Tarquinia: vicenda conservativa, restauri, tecnica di esecuzione.* (Firenze: Nardini Editore).

[14] Mura F, Cirigliano A, Bracciale MP and Rinaldi T 2020 Characterization of nanostructured calcium carbonate found in two ancient Etruscan tombs *AIP Conference Proceedings* **2257** 020011.

[15] Cirigliano A, Tomassetti MC, Di Pietro M, Mura F, Maneschi ML, Gentili MD, Cardazzo B, Arrighi C, Mazzoni C, Negri R and Rinaldi T 2018 Calcite moonmilk of microbial origin in the Etruscan Tomba degli Scudi in Tarquinia, Italy *Scientific reports* **8(1)** pp. 1-10.

[16] Cirigliano A, Mura F, Cecchini A, Tomassetti M, Maras D, Paola FD, Meriggi N, Cavalieri D, Negri R, Quagliariello A, Hallsworth J and Rinaldi T 2020 Active microbial ecosystem in Iron-Age tombs of the Etruscan civilization. *Environmental Microbiology* **23(7)** pp. 3957-3969.

[17] Dufresne WJB, Rufledt CJ and Marshall CP 2018 Raman spectroscopy of the eight natural carbonate minerals of calcite structure *Journal of Raman Spectroscopy* **49** pp. 1999-2007.

[18] Porto SPS., Giordmaine JA and Damen TC 1966 Depolarization of Raman Scattering in Calcite *Physical Review* **147(2)** pp. 608-611.

[19] Liu LG and Mernagh TP 1990 Phase transitions and Raman spectra of calcite at high pressures and room temperature. *American Mineralogist* **75** pp. 801-806.

3D X-ray Microscopy (XRM) investigation of exogenous materials inside mussels' organs

F Cognigni[1], S Dinarelli[1], M Girasole[2], G Longo[2], G Fabi[3] and M Rossi[1,4]

[1] Department of Basic and Applied Science, Sapienza University, Rome, Italy
[2] Institute for the Structure of Matter, National Research Council, Rome, Italy
[3] Institute for Biological Resources and Marine Biotechnology, National Research Council, Ancona, Italy
[4] CNIS - Center on Nanotechnology Applied to the Engineering of Sapienza, Sapienza University, Rome, Italy

E-mail: simone.dinarelli@uniroma1.it

Abstract. The diffusion of pollutants in the marine environment is nowadays a well-recognized issue that is attracting growing interest from the scientific and social communities. One of the possible strategies to study the effect of pollutants is to quantify their presence inside marine organisms that are directly exposed for a certain period to the polluted environment. Among them, mussels, commonly considered as "biological water filters", stand out as ideal candidates since they are stationary animals and their food intake comes only from the filtering of the surrounding water. Thus, the evaluation of the accumulation of exogenous pollutants, in particular high-density or metallic, inside the mussel's organs and specifically in its digestive glands, is of particular interest. In this paper we characterize the accumulation of exogenous materials in digestive glands of three different mussels by means of X-ray microscopy analysis. We provide evidence of how the unique capabilities of this technique allow reconstructing a full 3D image of an entire organ and how this image can provide valuable information to identify exogenous (non-biological) pollutants. Moreover, we take full advantage from the segmentation analysis of the images by discriminating different regions of the sample according to the density. With this experimental approach we measured the sizes of the exogenous pollutants and provided evidences that they accumulate preferentiality in the low-density regions of the organ, that are richer in ducts and secretive glands.

Key words: mussels, exogenous pollutants, X-ray microscopy

1. Introduction

The diffusion of pollutants, of different origins and composition, in the marine environment is gaining growing interest in the scientific community over the last years. The study of these contaminants and of their effects on the biota is paramount to define strategies to mitigate such effects. This is particularly true for high-density or metal-based pollutants which bears the highest harmful potential for ecosystems and biosystems. Among the many studies focused on these interactions, the investigation of samples collected directly from the marine environment are the most challenging, since the abundance and chemical composition of the pollutants are difficult to be defined [1, 2]. In the marine environment, the biological activity of pollutants depends on several factors, such as composition, chemical stability, structure, surface chemistry and presence of concomitant additional surfactant. In addition, the fate of pollutants inside a marine organism can be very different. A pollutant can be simply accumulated without generating harmful effects on the animal; it can induce biological modifications in cells, tissue or organs leading to the impairment of several functions; or can be neutralised and expelled from the organism itself. All these factors are affected by toxicity and accumulation rate at the level of tissue or entire organ [3,4]. Among the various organisms that can be studied, mussels (*Mytilus galloprovincialis*) are the ideal candidates for studies of pollution in marine environments since they are stationary animals and can be considered as natural water filters since their food intake comes from filtering the surrounding water. Nonetheless, mussels are edible organisms and can easily transfer the pollutants accumulated during their life through the food chain directly to humans. For this reason, the investigation of pollutants inside mussel's organs can be a reliable and direct way to monitor marine pollution. Several studies are available on the effect of pollutants on mussels [5,6], making use of different techniques, from biochemical characterizations [7-9], to spectroscopies [4] or even high-resolution microscopies such as Scanning Electron Microscopy (SEM) and Atomic Force Microscopy (AFM) [10]. Every given technique presents its own advantages and limitations in terms of balance between resolution, localization capabilities, quantity of material sampled and how much the sampled region can be representative of the entire animal or organ. A recently developed technique that holds the promise to be particularly useful in this field is X-ray Microscopy (XRM) that delivers a non-destructive characterization providing internal microstructural and compositional information about the sample and is particularly suitable for a multiscale and multimodal 2D/3D approach [11,12]. XRM is becoming increasingly popular thanks to its ability to reveal internal details of even large samples without the need of sectioning or dismantling, which can be a crucial advantage, especially in the case of biological samples. The experimental setup is typically composed of three main elements: (i) a source that produces the X-ray beam, (ii) a charge-coupled device (CCD) detector that collects the transmitted X-rays and (iii) a rotating sample stage, placed between the source and the detector, which firmly holds the sample in the desired position and ensures its controlled rotation during the analysis [13]. The contrast obtained in the final image strongly depends on the X-ray absorption phenomenon that occurs along the X-ray beam path (L) inside the sample, it is function of the attenuation coefficient value (x) in a certain position (dx) of the specimen and depends on the intensity of the X-ray beam before entering the sample (I_0). These parameters are related by the Beer-Lambert law:

$$I_L = I_0 e^{-Lxdx} \tag{1}$$

where I_L represents the transmitted X-ray beam intensity. Equation 1 describes the exponential decrease of a quasi-monochromatic X-ray beam passing through a heterogeneous object [13,14]. The transmitted X-ray beam reaches the detector where it is converted into visible light through a scintillator and the signal is then recorded and magnified by a camera. This is the process that leads to the acquisition of a single projection image and it is repeated iteratively over many different rotation angle values with respect to the X-ray beam direction, until a complete set of projection images is obtained [13,15]. A tomographic reconstruction algorithm is then used to obtain the 2D/3D reconstruction of the object from this set of projection images. Generally speaking, an X-ray Microscopy experiment can be run in two different modalities: absorption contrast tomography or phase contrast tomography. The first method allows to obtain the bi- and three-dimensional imaginary part of the refractive index distribution of the specimen, which is sensible to the X-ray amplitude changes. On the other hand, the second method is related to the change in wavelength of the X-ray beam along its path through the sample and measures the real part of the complex refractive index of the object. Although the phase contrast tomography is particularly suitable for low-Z materials such as polymers or biological samples that typically show a weak absorption [16,17], in this work the physical dimensions of the samples allowed us to work in absorption contrast tomography, thus being more sensible to the internal details of the structures. This work follows on the footsteps of previous works [9,10] that were devoted to the identification, with high resolution techniques, of nano and micro-sized pollutants in tiny sections of mussels. Through X-ray tomography, we can add quantitative and localized information of the exogenous pollutants, providing a micron scale map at the level of entire organs.

2. Materials and Methods

2.1. Sample collection and organs extraction

The mussels (*M. galloprovincialis*) analysed in this work were part of a more comprehensive study, whose complete set of information can be found in [9]. Briefly, the collected animals were rapidly transported in cold boxes in laboratory to perform the extraction of the digestive glands, which were subsequently dissected, cut in sections of 1 to 2 mm and then immersed in 2,5% glutaraldehyde phosphate buffer 0.2M solution with pH 7.2 and stored at 4°C until the resin embedding procedure.

2.2. Resin embedding procedure

The embedding procedure followed an already established protocol described in detail in [10]:

- 3x washes in Buffer Phosphate
- post-fixation in Osmium Tetroxide 1% for 8 hours
- 3x washes in Buffer Phosphate
- 2x exposure of 10 minutes each in solutions at growing concentration of Ethanol (from 25% up to 100% EtOH)
- 3x washes in Buffer Phosphate
- substitution of EtOH with a solution of 50% Propylene oxide for 10 minutes
- 2x passages in pure Propylene oxide solution for 20 minutes each

After this passage, the samples were cut in small pieces with a surgical knife and the resin embedding (Multilab 812 resin kit) was performed by exposing the samples to solutions with progressive concentration of resin increase in steps (from 33% to 100% resin).

The resin polymerization procedure (48 hours at 60°C) was performed after 1 hour from the exposure to pure resin, resulting in epoxy blocks, which were ready to be analysed with the 3D-Xray tomography without further processing.

2.3. X-ray Microscopy (XRM)

X-ray Microscopy analysis was performed using a laboratory X-ray microscope (Zeiss, Xradia Versa 610) at Sapienza Nanoscience & Nanotechnology Laboratories (SNN-Lab) of the Research Centre on Nanotechnology Applied to Engineering (CNIS) of Sapienza University of Rome. When scanned, the specimens were glued on a pencil lead and then fixed on the sample stage by the Pin Vise sample holder. The sample stage stands between the X-ray source and the detector and can rotate around a fixed axis to ensure a full 360° scan. For all the three samples the voltage and power of the X-ray beam were 50 kV and 4.5 W, respectively, the scans were performed from 0° to 360° using a 4x objective, acquiring 3201 projection images binned 2x2x2. The relative distances (sample-to-detector and sample-to-source) as well as the exposure time, were adjusted in such a way to optimise the signal-to-noise ratio and obtain a pixel size of 2 µm and then a voxel size of 8 μm^3. The total scanning time for the three samples changes accordingly to these parameters from around 9 hours up to 11 hours.

2.4. 3D models Reconstruction

3D models were reconstructed using the Reconstructor module of Zeiss Scout and Scan control software - Zeiss (Version 16.1.13038.43550) by which reconstruction parameters such as Centre Shift and Beam Hardening Constant were identified. At the end of the process a TIFF stack was obtained and subsequently imported in Dragonfly Pro software (Version 2020.1 Build 797, Object Research System) for post processing. Dragonfly Pro was used to (i) visualise the samples excluding the signal coming from the air around the specimens through the window levelling tab, (ii) generate the images presented in this paper and (iii) perform the segmentation procedure of the samples thus obtaining the sizes of the exogenous material. The subsequent data analysis of these sizes was performed with the Software OriginPro 2018.

3. Results and Discussions

Tomographies were collected from digestive gland samples from three different animals. The digestive glands are the organs responsible for the extraction of nutrients from the filtered water, essentially all the water that was filtered by the animal will pass through the ducts, glands and connective tissue that compose this organ. probably, most of the exogenous material that interacted with the tissue and was not expelled is stored in this organ, which is consequently expected to uptake an amount of pollutants greater than any other organ of the mussel. In figure 1, a complete tomography of a sample is reported.

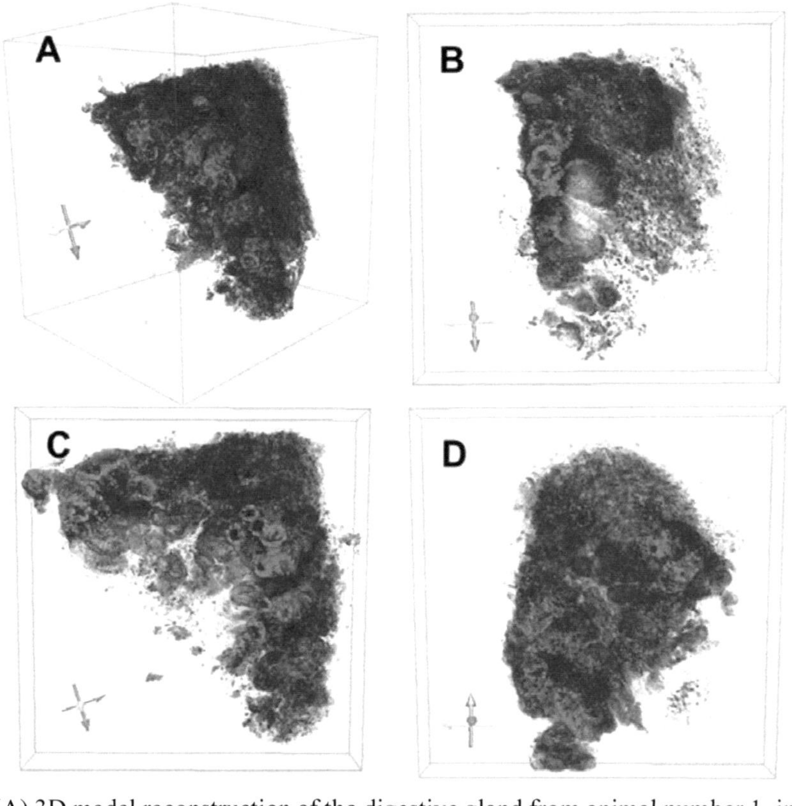

Figure 1. (A) 3D model reconstruction of the digestive gland from animal number 1, in which three thin sections have been extracted using a clipbox along the three orthogonal axes: (B) x-axis, (C) y-axis and (D) z-axis.

The 3D image shown in figure 1A allows to identify the presence of the various compartments of the tissue that are represented in greyscale, according to their density. To grant a different view of the sample, more similar to a standard slice as observed in optical microscopy, in figure 1B, 1C and 1D three virtual slices are reported. They were obtained by cutting the 3D model of the specimen along the x-y, y-z and z-x planes, respectively. These thin sections appear very similar to optical microscopy images of the histological sections, but with a higher resolution and without the need of sample staining. While the 3D X-ray tomography is not capable of identifying the elements that compose a complex sample such as a biological organ, it allows discriminating the different densities inside the volume. This allows the identification of the presence of exogenous materials by looking for elements with much higher densities that, in an organic matrix, are not expected. This procedure is called segmentation and it allows the complete identification of the exogenous material in the three samples analysed in this work (figure 2). By dramatically reducing the opacity of the sample reconstruction (as seen comparing the left and right images), we are able to visualize and localize the high-density pollutants (red).

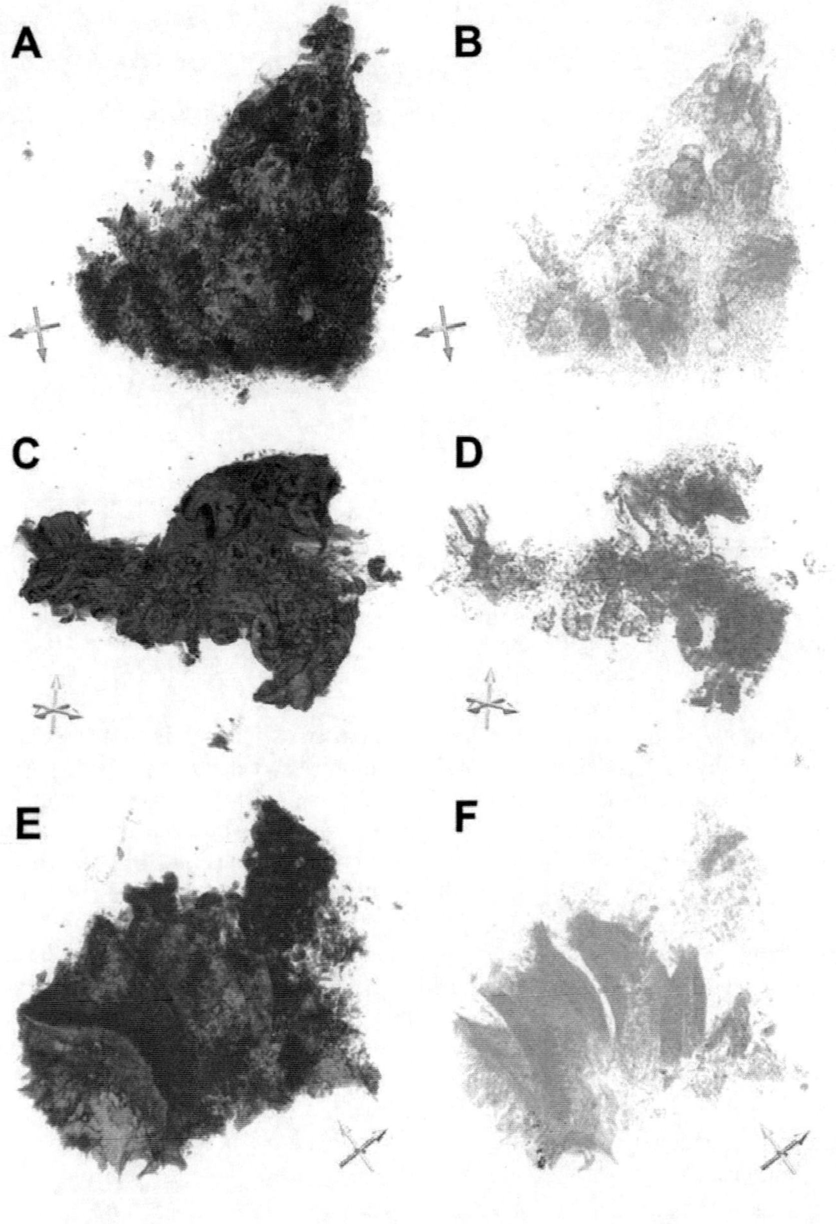

Figure 2. Exogenous pollutants (red) were segmented to highlight their distribution over the whole sample volume after reducing the object's opacity. (A-B) digestive gland from animal 1, (C-D) digestive gland from animal 2 and (E-F) digestive gland from animal 3. There is a preferential distribution of exogenous pollutants inside regions rich in channels and tubules with respect to the connective tissue. A complete segmentation result is shown in figure 3 for the digestive glands of the animal number 2, where the different regions of the organ are mapped in different colours, according to their density.

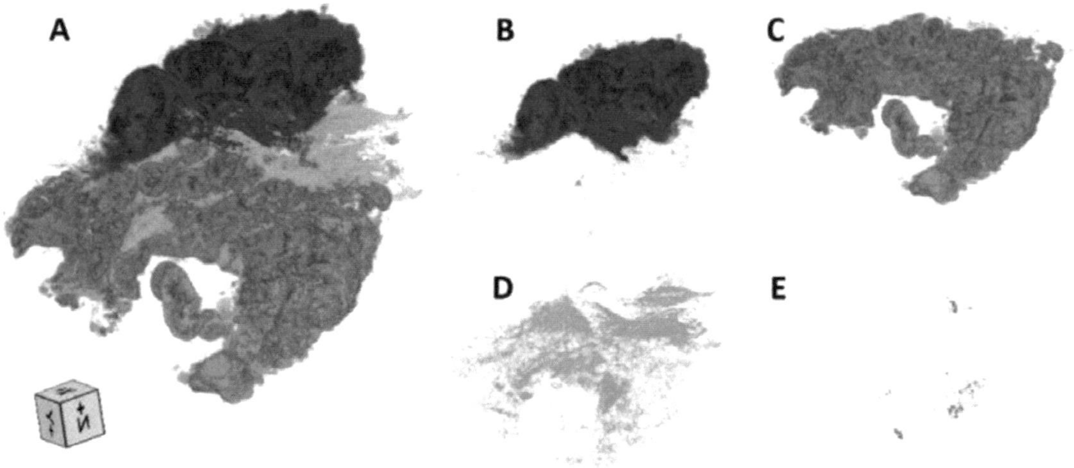

Figure 3. (A) 3D model reconstruction of the digestive gland from animal number 2, where various phases of the organ such as connective tissue, cellular compartment and exogeneous pollutants were segmented. Then, every phase has been extracted and reported separately: (B) high density region rich in tubules and channels, (C) soft density region rich in tubules and channels, (D) connective tissue and (E) exogenous high-density pollutants.

More in detail, in blue (Figure 3B) is shown a high-density region rich in tubules and channels, in green (Figure 3C) is shown a low-density region rich in tubules and channels, in yellow (Figure 3D) is shown a region of connective tissue while in red (Figure 3E) are shown the exogenous materials which exhibit a very high density. From these figures some consideration can be done: (i) the connective tissue connects the two regions of tubules and channels at different densities and (ii) there is a significant presence of exogenous materials that is principally localised in the lower density region. The spatial localization of the technique allows determining that the exogenous materials are not uniformly distributed in the organ; they are not present on the surface of the organ but only inside the channels and tubules (Figure 3A). From these images, it is immediately clear how the segmentation procedure easily and automatically identifies different biological components in the tissue as well as evidencing the exogenous materials. These considerations allow to determine that in all the three organs there is a significant presence of exogenous material that is preferentially localised in the regions rich in channels and tubules, while there is practically no pollutant presence in the connective tissues.
To deal with the quantification of the presence of exogenous materials in the samples, we report in figure 4 their size distributions.

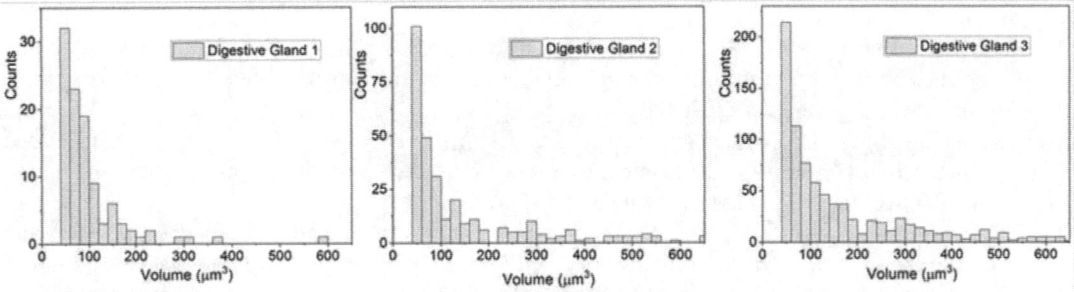

Figure 4. Histograms of the values of the volume of each exogenous material measured in the three samples analysed in this work.

In these histograms are reported the volumes of each single contaminant measured through the segmentation procedures described above. It should be noted that, to comply with the instrumental minimum voxel size of 8 μm^3 and with the necessity of having at least 5 voxels to really identify a single particle, all the contaminants with volume smaller than 40 μm^3 were considered not significant and therefore were removed from the histograms. The shape of the three size distributions appears similar, with a very high peak at lower volumes coupled with the presence of several larger contaminants. The major difference among the samples resides in the total number of contaminants measured, which can depend on several parameters such as the intrinsic different nature of the contamination encountered by the three animals but also by the size of the organ section analysed (i.e., the total volume sampled).

To roughly estimate the sensitivity of our approach we measured the relative volume occupied by the exogenous material, defined by the ratio between the total volume occupied by the exogenous material and the total volume sampled. By this simple calculation we find that this ratio is around 0,004% for sample 1, 0,02% for sample 2 and 0,09% for sample 3. It means that even if the exogenous materials visualized in figures 2 and 3 were widely distributed and clearly visible, their quantity with respect to the entire organ is extremely low, thus highlighting the great sensitivity of this technique, capable to identify and localize highly dispersed components inside very complex and structured specimens.

4. Conclusions

We provided evidence that the 3D X-ray tomography can be a valid tool to identify and localise exogenous materials inside biological specimens, with a micrometric resolution. Such technique can complement other high-resolution approaches such as SEM, TEM and AFM, by providing information on the localization and distribution of contaminants at the level of an entire organ. The key point is the unique capabilities of this technique to provide information on the internal structure of the sample under study, generating a real 3D map of the specimen that can be post-processed by slicing the sample at the desired position focussing directly on the structures of interest. We also demonstrate how the segmentation procedure can be a very powerful tool in discriminating different regions of the sample according to the local density. While this work is not a comprehensive study of the meso-pollutant inside mussel's organ, but a proof of concept of the capabilities of the technique, it shows the capabilities of XRM and its importance in conjunction with other analysis tools. For instance, 3D X-ray tomography could be a first step in a high-resolution approach in the study of contaminants in biological specimens, educating the investigation towards the tissue areas, which could provide the most interesting results using high-resolution or chemically resolved techniques. Once a region of interest packed with exogenous pollutant is observed, correlative microscopy workflows can be exploited to enrich our knowledge with chemical and compositional information down to nanometre scale.

Acknowledgments

This research has been realized with the financial support of the project ATOM (Advanced Tomography and Microscopies), granted by Lazio Region (call "Open infrastructures for research").

References

[1] Griffitt RJ, Luo J, Gao J, Bonzongo JC and Barber DS 2008 Effects of particle composition and species on toxicity of metallic nanomaterials in aquatic organisms *Environmental Toxicology and Chemistry* **27** pp. 1972-1978.

[2] Tangaa SR, Selck H, Winther-Nielsen M and Khan FR 2016 Trophic transfer of metal-based nanoparticles in aquatic environments: a review and recommendations for future research focus *Environmental. Science Nano* **3** pp. 966-981.

[3] Viarengo A and Nott JA 1993 Mechanisms of heavy metal cation homeostasis in marine invertebrates *Comparative Biochemistry and Physiology Part C: Toxicology* **104** pp. 355-372.

[4] Viarengo A, Ponzano E, Dondero F and Fabbri R. 1997 A simple spectrophotometric method for metallothionein evaluation in marine organisms: an application to Mediterranean and Antarctic molluscs *Marine Environmental Research* **44** pp. 69-84.

[5] Rocha TL, Gomes T, Sousa VS, Mestre NC and Bebianno MJ 2015 Ecotoxicological impact of engineered nanomaterials in bivalve molluscs: an overview *Marine Environmental Research* **111** pp. 74-88.

[6] Sousa JC, Ribeiro AR, Barbosa MO, Pereira MFR and Silva AM 2018 A review on environmental monitoring of water organic pollutants identified by EU guidelines *Journal of Hazardous Materialsu* **344** pp. 146-162.

[7] Cravo A, Lopes B, Serafim Â, Company R, Barreira L, Gomes T and Bebbiano MJ 2009 A multibiomarker approach in Mytilus galloprovincialis to assess environmental quality *Journal of environmental Monitoring* **11** pp. 1673-1686.

[8] Gomiero A, Da Ros L, Nasci C, Meneghetti F, Spagnolo A and Fabi G 2011 Integrated use of biomarkers in the mussel Mytilus galloprovincialis for assessing off-shore gas platforms in the Adriatic Sea: results of a two year biomonitoring program *Marine Pollution Bulletin* **62** pp. 2483-2495.

[9] Gomiero A, Spagnolo A, De Biasi A, Kozinkova L, Polidori P, Punzo E, Santelli A, Strafella P, Girasole M, Dinarelli S, Viarengo A, Negri A, Nasci C and Fabi G 2013 Development of an integrated chemical, biological and ecological approach for impact assessment of Mediterranean offshore gas platforms. *Chemistry and Ecology* **29** pp. 620-634.

[10] Dinarelli S, Longo G, Cannata S, Bernardini S, Gomiero A, Fabi G and Girasole M 2020 Metal-based micro and nanosized pollutant in marine organisms: What can we learn from a combined atomic force microscopy-scanning electron microscopy study *Journal of Molecular Recognition* **e2851**.

[11] Duncan KE, Czymmek KJ, Jiang N, Thies AC and Topp CN 2022 X-ray microscopy enables multiscale high-resolution 3D imaging of plant, cells, tissues, and organs *Plant Physiology* **188** pp. 831–845.

[12] Dini D, Cognigni F, Passeri D, Scaramuzzo FA, Pasquali M and Rossi M 2021 Review—Multiscale Characterization of Li-Ion Batteries through the Combined Use of Atomic Force Microscopy and X-ray Microscopy and Considerations for a Correlative Analysis of the Reviewed Data *Journal of The Electrochemical Society* **168** pp. 126522

[13] Carmignato S, Dewulf W and Leach R 2018 Industrial X-Ray Computed Tomography *Springer International Publishing.*

[14] Buzug T and Thorsten M 2008 Computed Tomography: from Photon Statistics to Modern Cone-Beam CT *Springer International Publishing.*

[15] Stock SR 2020 MicroComputed Tomography - Methodology and Applications *CRC Press.*

[16] Reichardt M, Frohn J, Khan A, Alves F and Salditt T 2020 Multi-scale X-ray phase-contrast tomography of murine heart tissue *Biomedical Optics Express* **11** pp. 2633-2651.

[17] Garcea SC, Wang Y and Withers PJ 2018 X-ray computed tomography of polymer composites *Composites Science and Technology* **156** pp. 305-319.

AUTHOR INDEX